# The City Project

Strategies for Smart and Wise Sustainable Urban Design

## Volume 9

The book series The City Project reports on applied research and operational developments that promote urban renewal and the sustainable transformation of contemporary cities. Inspired by the "City of Man" as imagined by Adriano Olivetti and Ernesto Nathan Rogers, and going beyond the concept of the smart city and related technological advances, the series' goal is to present holistic, practice-oriented and multidisciplinary strategies for realizing the City 4.0, i.e., the city of the fourth industrial revolution, in keeping with the objectives of the UN's 2030 Agenda for Sustainable Development.

In particular, the series reports on effective design, planning and management approaches that leverage urban and architectural design skills, engineering, environmental and social expertise, and administrative abilities alike. It welcomes books on each of the aspects mentioned above, as well as studies analyzing multiple aspects, their interactions and/or holistic solutions. The City Project addresses a very broad readership, including designers, engineers, architects, social scientists, stakeholders and public administrators, who deal with various aspects of the realization of the City 4.0. It publishes theoretical investigations into the contemporary built environment, international case studies, and pilot projects concerning urban renewal and the regeneration of urban areas, as well as the proceedings of key international conferences.

Books published in this series are devoted to supporting education, professional training and public administration. Outstanding PhD theses on emerging topics, if properly reworked, may also be considered for publication. The series is published with the support of the Smart City 4.0 Sustainable LAB, an interdisciplinary teaching and research project on future cities initiated by the University of Parma, and jointly implemented with other regional universities (the University of Bologna, University of Ferrara, and University of Modena and Reggio Emilia).

**About the Cover**

The cover of the book series The City Project features a painting by Carlo Mattioli (C. Mattioli, *Estate in Versilia* , 1974, oil on canvas cm. 118 X 70, Catalog n. 1974D0029, Courtesy of Fondazione Carlo Mattioli, thanks to Anna Zaniboni Mattioli)

*The horizon of poppies painted by Carlo Mattioli between the dark background of the forest and the white plane of the wheat, becomes for us, thanks to a transfiguration of meaning that aligns with the attitude towards abstraction rooted in the figure of the painter, a city which is intertwined with its landscape, evoking the idea and the possibility of recomposing a balance and seeking an integration between settlement and environment, between human space and natural element.*

Antonio D'Aloia • Simone Scagliarini
Editors

# Law and Rights in the Smart City: The Challenge of Regulation

Springer

*Editors*
Antonio D'Aloia
Department of Law, Politics and International Studies
University of Parma
Parma, Parma, Italy

Simone Scagliarini
Department of Economics "Marco Biagi"
University of Modena and Reggio Emilia
Modena, Italy

ISSN 2730-6992 ISSN 2730-700X (electronic)
The City Project
Strategies for Smart and Wise Sustainable Urban Design
ISBN 978-3-032-13070-9 ISBN 978-3-032-13071-6 (eBook)
https://doi.org/10.1007/978-3-032-13071-6



This Springer imprint is published by the registered company Springer Nature Switzerland AG
The registered company address is: Gewerbestrasse 11, 6330 Cham, Switzerland

# Acknowledgments

*We would like to thank Dr. Mario Mendillo (PhD student, University of Parma) for his valuable work in formal editing of this volume and management of the publication process.*

# Contents

# Law and Rights in the Smart City: *Introductory Remarks*

Antonio D'Aloia and Simone Scagliarini

The Conference from which this Volume is drawn falls within the framework of the *SMART CITY 4.0 Sustainable LAB* project in which the universities of Emilia-Romagna participate.

This involves a dialogue not only between Universities but also between disciplines, albeit very distant from each other, which contribute to analyze, from a design perspective, the profile of the *smart city*, which precisely presupposes a (physiological) interdisciplinary approach motivated by the breadth and transversality of the elements and subjects that affects. Such an approach is also justified by reason of the intrinsic delay that legal reflection suffers in relation to scientific progress and technological experimentation: a delay that is reflected in a scenario where law finds itself having to almost "chase after" technique and science, with the effect of arriving at categorizations and interpretative readings that—once consolidated—risk to appear already outdated. The integration between different knowledge and competencies, therefore, becomes the keystone, the (necessary) prerequisite for an effective regulation: and such integration likely represents the only way to approach the discourse on *smart cities,* effectively and—as far as possible—comprehensively.

It must also be considered that—beyond a necessary (administrative) activity of regulating new urban systems—the *smart city* model inevitably affects the sphere of constitutionally protected rights.

Consider, just as an example, the relationship between center and periphery, between large urban agglomerations and small villages or centers in internal areas; the use of ICT tools (and all that follows) and the utilization of artificial intelligence; privacy protection; the necessary rethinking (or perhaps the *ex novo* conception) of innovative mechanisms of public participation; the public services system and the guarantee of social rights; environmental protection, or rather the (indispensable) integration of this primary objective in all decision-making processes, at all levels of governance; the very functioning of elective assemblies.

As Giovanni Flick explains very well (in his *Elogio della città? Dal luogo delle paure alla comunità della gioia*, Milan, 2019), the City is not simply one of the many segments of the local power system that we know to be extremely articulated and complex (the Republic of autonomies regulated by articles 114 et seq. of our Constitution), but is—in all respects— *"a social formation, one of those places where the*

A. D'Aloia (✉)
Department of Law, Politics and International Studies, University of Parma, Parma, Parma, Italy
e-mail: antonio.daloia@unipr.it

S. Scagliarini
Department of Economics "Marco Biagi", University of Modena and Reggio Emilia, Modena, Italy
e-mail: simone.scagliarini@unimore.it



A. D'Aloia, S. Scagliarini (eds.), *Law and Rights in the Smart City: The Challenge of Regulation*, The City Project 9, https://doi.org/10.1007/978-3-032-13071-6_1

*reflection, the refraction of the great themes of contemporary constitutionalism is realized*".

In this sense, then, 'smart' risks being a partial descriptor, especially if interpreted only in the sense of technological efficiency, of the ability to function, thus looking at only one side of the problem, and perhaps not even the most important one.

The real challenge is a city capable of holding together its contradictions, its conflicts, of not leaving anyone behind, of offering a space for dialogue and democratic construction.

Correctly then, scholars like Stefano Zamagni, or Derrick De Kerckhove, prefer to speak of (and seek) Wise City, or Care City: a wise, inclusive, sustainable City, capable of reflecting and amplifying the constitutional value of solidarity, and of being a space where the removal (or at least the pursuit of removal) of those economic and social obstacles that prevent substantial equality is concretely possible. Smart, therefore, as an instrument and not as an end, according to Fabrizio Fracchia's vision, to "improve the quality of life of citizens and businesses" (F. Fracchia, *Smart City: share to innovate (with the risk of excluding?)*, in Federalismi.it, n. 22/2015, 10, 17 ff., 25).

Taking into consideration these synthetic observations, the debate that took place during the Conference, while following the *fil rouge* represented by the focus on *smart city* regulation, then sees the combination of two macro-themes: rights in the smart city and law of the smart city.

The presentations of the first session (Olivito, Gragnoli, Rossi Chauvenet and Buzzetti) focus on the topic of rights in the smart city, that is, on the impact that urban transformation in a *smart* sense can have on constitutional rights. In particular, Olivito signals the need for the *smart* city to be *constitutionally oriented*, which means that the technological implementation process must be guided by public authorities to ensure its function of promoting the person and removing inequalities, with the parallel objective of preventing this evolution of the city concept from becoming an opportunity to reproduce, if not amplify, existing inequalities. The risk that is glimpsed is indeed that of not being able to direct development (in general, technological and of the *smart city* model) toward socially desirable objectives, with consequent increase in the rate of social divergence and marginalization phenomena of complex and uncertain resolution.

It will therefore be necessary that digital innovation, and the re-modeling of the environments in which we live in function of the use of new technologies, will be accompanied by the elaboration of careful social inclusion policies, moving from the postulate that without human coexistence the city does not exist and that the latter—far from being a mere spatial datum—represents a true socio-cultural entity, made of interactions and relationships between citizens, whose improvement corresponds to the improvement of the functioning of the urban entity itself.

Gragnoli, for his part, reflects on the transformation of work in recent years, in the sense of the shift toward agile (*smart*) forms, in which the performance takes place in homes that however were not conceived and designed to be (also) workplaces.

The theme then arises of thinking about the *smart city* also from the perspective of becoming a place suitable for a *smart* transformation of work as well.

Moreover, as the recent and dramatic pandemic event demonstrated, the proper functioning of labor markets and social protection systems—especially in a world that changes so quickly, making it extremely difficult for the legislator to prepare in rapid times the political-legal instruments adequate to respond to new needs—represents the premise of the European discourse on equity and social rights (as also highlighted by the European Commission itself, already with Recommendation n. 761 of April 26, 2017).

Rossi Chauvenet and Buzzetti then focus on the inevitable data sharing required by the city's *smart* transformation, with the problems that emerge to ensure compliance with GDPR prescriptions.

Moreover, the legislation on the free circulation of personal data and the discipline of the *Data Governance Act* seem to offer important instruments for the development of public services capable of making cities more intelligent. On the other hand, some experiences in the field of mobility have already been successfully implemented.

The presentations of the second session (Gaspari, Raffiotta, Gastaldo, Al Mureden) instead focus on the *law of* the smart city, placing the focus on *governance*.

Gaspari, whose reflection serves as a hinge between the first and second sessions, starts from the assumption that, if a right to the city exists, and the city, as previous presentations have demonstrated, is not a neutral non-place but the space within which citizenship is exercised, then the theme arises of how to guarantee citizen participation.

A problem that emerges both *ex ante*, in the design and transformation phase of the city, ensuring a *bottom-up* approach that incorporates *by design*, within itself and from the origin, direct citizen involvement, and *ex post*, when a *smart* city offers more channels and new forms of participation, relaunching and developing the use of *e-government*.

Indeed, the *intelligence of the city* can be said to be achievable only if it passes through the contextual valorization and promotion—beyond digital infrastructures—also and above all of human capital, which expresses its potential in the city and which can, in this way and hopefully, lead to economic and social progress that can be defined as concretely *sustainable* (according to the tripartite declination of the concept of sustainability, which affects the social element as well as the economic and environmental ones).

And furthermore, it should be considered that the democratic involvement of the citizen can limit the emergence of a further risk, namely that related to the possibility that, to the increase in the quantity of shared data, there may correspond a stiffening of control systems, an exasperation of surveillance activities, suitable to strongly (and negatively) affect the individual legal sphere of the citizen.

To this last point connects Raffiotta's presentation, which raises the question of how a smart city is governed.

The legislation on the *governance* of local entities, even those of more recent establishment and with a generic direction toward technological evolution, such as metropolitan cities, does not seem adequate to face the new challenges.

The theme then arises of how and to whom it falls to imagine new solutions that allow evolution, if deemed desirable, in this direction, where the role of the State, also to ensure that inclusive purpose of which we have spoken, seems irreplaceable both for the enormous investments that are needed, for which our Country already appears to be behind, and for the current inability of many Municipalities (especially smaller ones) to give life even to the simplest forms of service digitalization.

Precisely to these small realities was dedicated Gastaldo's intervention, which emphasizes how minor Municipalities (for example mountain ones) cannot be abandoned to themselves, accentuating a *digital divide* between territories, as well as between citizens, but it is instead necessary to imagine forms of connection, starting from the awareness of the problem; awareness that also seems to be found in the NRRP.

The renewal of the urban model cannot disregard the fact that all territorial components (vast urbanized areas, intermediate and minor cities, mountain municipalities and internal areas, but also sections and settlements internal to the same urban area) enter into correlation and dialogue with each other, integrating and generating a relational system (a *system of systems*) able to operate jointly, ensuring the involvement and proper functioning of all its gears.

Finally, Al Mureden's presentation focuses on intelligent mobility and—in particular—on autonomous driving, as a case study of the impact of technological innovation on the urban context (moreover with very challenging ethical questions: just think of the theme of responsibility for possible accidents caused by the behavior of driverless cars), such as to require urban redesign from the perspective of *smart mobility*.

The reflections that emerged from the debate are rich and offer a very clear picture of the complexity of the theme and the necessity of an interdisciplinary approach to the question: to this debate the Conference wanted to give an active contribution.

# The "Smart City": An Instrument that Tackles Inequalities or that Amplifies them? The Constitutional Perspective of Social Peripheries

Elisa Olivito

**Abstract**

The importance of the smart city as a concept—both in its technological and environmental declination—finds final confirmation in the NRRP Mission 5c2.2 on "*Urban regeneration and social housing*," which marks the relationship between urban peripheries and the smart city. The spread of this new and multi-purpose notion leads to debates that impact first and foremost on the constitutional level. Before thinking of smart solutions for cities, it is crucial to discuss what kind of smart city that is to be built. From a constitutional perspective it must be a city inhabited by embodied people, who consider things as goods before commodities and in it entertain meaningful human relationships and not only economic transactions. For this purpose, those who look to the smart city from the perspective of constitutional law are tasked with problematizing this notion in light of the Constitution.

E. Olivito (✉)
Dipartimento di studi giuridici ed economici, Università di Roma "La Sapienza", Rome, Italy
e-mail: elisa.olivito@uniroma1.it

## 1 The Fortunes of an Intangible Concept: Some Terminological Premises

In Italy, one may note the growing use of the label "smart city," especially at the level of soft law, good practices, and plans of action, as well as its intertwining with technological development and with environmental issues.

The fortunes of the smart city concept only found final confirmation in the National Recovery and Resilience Plan (*Piano Nazionale di Ripresa e Resilienza*—hereinafter NRRP), which abounds in smartness (smart procurement, smart working, smart agriculture, smart grids, smart road network, smart stations, smart cities).

In particular, *Mission 5c2.2* on "Urban regeneration and social housing," calls for two types of investment, marking the interrelationship between urban peripheries and the smart city.

On the one hand, there are investments in urban regeneration (2.1.). They are aimed at providing Municipalities that have populations exceeding 15,000 inhabitants with contributions for investments in urban regeneration, to reduce situations of social decay and marginalisation and to improve the quality of urban decor and of the social and environmental setting.


A. D'Aloia, S. Scagliarini (eds.), *Law and Rights in the Smart City: The Challenge of Regulation*, The City Project 9, https://doi.org/10.1007/978-3-032-13071-6_2

On the other hand, and in parallel with urban regeneration, an investment is planned in the Integrated Urban Plans (2.2.). It is dedicated to the peripheries of the metropolitan cities and is aimed at developing a participatory urban planning. The goal is that of transforming vulnerable territories into sustainable smart cities, and of limiting the consumption of developable land. This investment contemplates the development of participatory urban planning programmes for urban regeneration, which are aimed at the improvement of large decayed urban areas, as well at regeneration, and at economic revitalisation. Particular attention is devoted to creating new services for individuals and bettering the accessibility of infrastructures, also with the purpose of transforming vulnerable metropolitan territories into "*efficient, sustainable, and productive territories, increasing their value where possible*" (emphasis added).

While the NRRP provides no explanation of the concept of smart city, the EU Commission's definition is rather vague:

> [a] smart city is a place where traditional networks and services are made more efficient with the use of digital solutions for the benefit of its inhabitants and business. A smart city goes beyond the use of digital technologies for better resource use and less emissions. It means smarter urban transport networks, upgraded water supply and waste disposal facilities and more efficient ways to light and heat buildings. It also means a more interactive and responsive city administration, safer public spaces and meeting the needs of an ageing population.[1]

One might cite a great many, and no less generic, references to the the smart city concept, as it was now the latest but empty buzzword in smart urbanism: empty because it has to be filled with meaning and contents, without which every finding remains suspended, and many questions arise. Moreover, from the legal standpoint, the vagueness of the smart city concept (which is different from that of the "smart community"[2]) corresponds to the vagueness of the very notion of "city". It has become an object of reflection also of constitutional law scholars, who are in fact more accustomed to reasoning about the concept of "municipality" understood as a territorial public body.[3]

The perplexities over the indiscriminate use, for the most diverse purposes, of so alluring a concept are linked above all to the anomaly of its origin. Indeed, beyond being employed as a mere slogan, the concept first emerged between the late 1990s and the early 2000s in the statements of multinational corporations such as IBM, Cisco, and Siemens, which started to develop management technological solutions for infrastructures and urban services. Only later did it attract the attention both of politics, with the related national and European financing programmes, and in academia, which began to raise questions about it.[4]

In spite of the indeterminacy and the lack of a shared definition (or perhaps precisely because of this),[5] the smart city concept produced debates among scholars that took on the characteristics of either/or and of clear opposition between sceptics and supporters.

On the one hand, the supporters of the smart city believe that this concept has a potential for positive transformation of urban areas, and the measures that it conveys might have beneficial impacts on cities in terms of safety, efficiency, and sustainability; on the other, the sceptics emphasise the smart city's criticalities and nega-

[1] EU Commission, *What are smart cities?*, at: *commission.europa.eu/eu-regional-and-urban-development/topics/cities-and-urban-development/city-initiatives/smart-cities_en* (last visited 25-05-2025).

[2] See Delle Cave, *«Comunità intelligenti», enti locali, mobilità sostenibile: le* smart city *al cospetto del potere pubblico*, in P. Stella Richter (ed), *Ripensare la città e il suo, Milano*, Giuffrè Francis Lefebvre, 2022, 369 ff.

[3] As I argued in E. Olivito, *(Dis)eguaglianza, città e periferie sociali: la prospettiva costituzionale*, in *Rivista Aic*, 1/2020, 4 ff.

[4] On this irregular origin, see A. Vanolo, *Smartmentality: The Smart City as Disciplinary Strategy*, in *Urban Studies*, 51(5)/2014, 888 ff.

[5] Cf. M. Ziosi, B. Hewitt, P. Juneja, M. Taddeo, L. Floridi, *Smart Cities: Reviewing the Debate About their Ethical Implications*, in *AI & Soc*, online, 30 September 2022 (last visited 25-05-2025).

tive repercussions, for example from the standpoint of profiling and breach of privacy, the commodification of human experience, technology inequalities, and social and spatial marginalisation.[6]

The choice between one option and the other cannot be made *a priori*, since the option for the enthusiastic reading or for the sceptical one depends on the contents ascribed to the smart city as a container, as well as on the implementation of the concept and its intertwining with social justice issues. Therefore, the smart city's conceptual ambivalences, emphasised even by those who do not consider themselves entirely sceptical about them, cannot be ignored. We are referring here to the risk that the concept might convey a technocratic governance of the city by enterprises oriented towards the surveillance of citizens and data monitoring. It leads to an underestimation of the specific urban contexts and on the sterilisation of the needs of those who inhabit the city.[7]

## 1.1 The Vagueness of the Debate About the "Smart City": What Are We Speaking About?

The opposing interpretations of the smart city raise the question as to whether the concept's ambivalence is a consequence of the difficulty of circumscribing it, or, vice versa, whether the concept's indeterminacy is instrumental to the ambiguity of the purposes underlying it.

It is therefore important to understand whether the inconsistencies underlying the smart city concept is connatural to it, and whether they can be overcome.

Although some believe a distinction can be drawn between a concept of smart city that relates more to corporate and financial needs, and another one which, preferably, underlies a broader social and economic dimension of the city,[8] the fact remains that this distinction is a hope or thesis more than it is a reality. The idea that the smart city can be the place or the conceptual paradigm within which the two perspectives flow beneficially together remains to be demonstrated, on both the theoretical and the practical level.

In fact, the attempts to make the smart city a tangible reality contain a host of paradoxes and contradictions. They derive in the first place from the difficulty of reconciling an urban intelligence shaped and guided by the global interconnections of financial capitalism with the specific features, needs, and experiences of the urban settings upon which this intelligence is to be placed.

In this sense, the smart city paradigm proposes objectives that can by all means be agreed to, first and foremost from the standpoint of environmental sustainability, but under a twofold condition: *i.* attention to these targets must not be an excuse for losing sight of the multiple forms of socioeconomic inequality, which find fertile ground above all in contemporary cities[9]; *ii.* we must rid ourselves of the presumption that the smart city model can be replicated regardless of the diversity and specific features of urban settings and territories, because this universalist presumption is not only wholly fallacious, but is also misleading.

In reality, the narrative of the smart city as a place or tool capable of bringing together interests for economic competitiveness, technology, and digitalisation with attention to human capital (individual and collective) and to environmental resources cannot obscure a far more problematic reality than the reconciled intersection between the two plans.[10] In this regard, there is no hiding that the aspiration to develop the smart city and

[6]On this debate, see A. Ferreira, F.P. Oliveira, K. C. von Schönfeld, *Planning Cities Beyond Digital Colonisation? Insights from The Periphery*, in *Land Use Policy*, 114/2022, 2 ff.

[7]As emphasised by M. Angelidou, *Shortcomings to Smart City Planning and Development: Exploring Patterns and Relationships*, in *Tema - Journal of Land Use, Mobility and Environment*, 10(1)/2017, 79 ff.

[8]See L. Sartori, *Alla ricerca della* smart citizenship, in *Istituzioni del federalismo*, 4/2015, 928.

[9]Cf. E. Swyngedouw, M. Kaika, *Re-naturing cities: great promises, deadlock . . . and new beginnings?*, in K. Archer, K. Bezdecny (eds), *Handbook of Cities and the Environment*, Elgar, 2016, 50 ff.

[10]See S. Zuboff, *The Age of Surveillance Capitalism. The Fight for a Human Future at the New Frontier of Power*, New York, Public Affairs, 2019.

the language to express it are the product of a functionalist and technocratic idea of the city, which thus becomes subservient to the needs of digitalised financial capitalism and its players.[11]

From this standpoint, the smart city is an example of the politically ambiguous use of the city as a synecdoche, in which a part (urban digital innovation) is taken for the whole (the city as a homogeneous entity). The contradictions, the redistributive issues, and the different (not only economic) standings within the physical and legal space of the city are thus obscured by the apparently merely descriptive use of smart city concept. In this way, "what is good for one (generally the business community) is good for all,"[12] since

> [t]hese views of 'the city' support a search for consensus politics, in which the claims of minority or powerless or disenfranchised or non-mainstream groups are considered disturbing factors in the quest for policies benefiting 'the whole'.[13]

In the alternative between a narrative hinging upon the real innovations underlying the "smart city" and an adaptively conservative representation, reflection upon the speculative potentials of the concept requires not only that a perspective be assumed, but also that questions be raised on the innovations that the smart city brings.

On the one hand, in fact, the smart city concept, by its indefinite nature, risks being translated into an empty container of urban digitalisation. Through it, capitalism only expands, thinking up new needs, new forms of consumption and offering individuals new products and new services for urban life: consider digital platforms like *Airbnb*, whose logic of short-term profit is part of a scenario in which the ability for public powers to regulate the phenomenon is limited. In this case, the regulation or lack of regulation of the markets connected to the sharing economy is one of the factors that can reject or reinforce new forms of gentrification connected to tourism and the related creation of "social peripheries".[14]

On the other hand, the smart city is conceived as a synthetic formula of technological solutions that large enterprises place at the service of city administrations, such as for example the systems to manage large infrastructures and urban services more generally. In this sense, it is a formula that denotes relevance to urban policies.

As for this second dimension of the smart city, problems in terms of the link between the public and private spheres have been precisely observed, as it is wholly clear that, in this way, the supply and management of infrastructures and public utilities by private parties becomes not only attractive, but above all a source of considerable profit.

In fact, while multinational corporations offer to design technologies that allow water or electrical utilities to be managed efficiently (which means more affordably and conveniently from the functional standpoint), the opportunities for savings and improved efficiency raise conflicts on the political level. And these conflicts are bound to intensify in a setting of already considerably reduced public spending.

Moreover, a further problem is raised, linked to the spontaneous phenomenon known as "lock-in":

> once we are bound to a technological solution, it is hard to change trajectory. If we begin to use the system produced by any multinational enterprise and the workers, managers, or users grow accustomed to this system, it is very hard to break away from it. This means we are tied for very long periods of time to services, maintenance, training, and everything else that revolves around that product or technology, resulting in situations of near monopoly.[15]

---

[11] See D. Harvey, *From Managerialism to Entrepreneurialism: The Transformation in Urban Governance in Late Capitalism*, in *Geografiska Annaler*, 1/1989, 4; P. Marcuse, *The Three Historic Currents of City Planning*, in G. Bridge, S. Watson (eds), *The New Blackwell Companion to the City*, Malden, 2011, 643 ff.

[12] P. Marcuse, *'The city' as perverse metaphor*, in *City*, 2/2005, 248.

[13] P. Marcuse, *'The city' as perverse metaphor*, in *City*, 2/2005, 252.

[14] I emphasised this aspect in E. Olivito, *(Dis)eguaglianza, città e periferie sociali: la prospettiva costituzionale*, 46 ff. supra note 3.

Regarding the concept of social periphery, see below.

[15] A. Vanolo, *«Non puoi non desiderare di essere intelligente!». Le Smart City e il futuro digitalizzato delle nostre vite. Intervista di Martina Mancinelli*, at *www.malorarivista.it*, 17 June 2022 (last visited 05-05-2023, my translation).

## 2 Claiming the Autonomy of the Constitutional Perspective Regarding the Smart City: The Question of Method, Above All, Together with a Non-neutral Idea of Constitutional Law

Beyond the easy multidisciplinary or interdisciplinary rhetoric, in examining the smart city as a heuristic category, it is decisive first of all to ask—and to ask ourselves—a fundamental methodological question, in order to clarify where we stand.

In terms of method, it is important to observe the distance from that type of legal realism so in vogue today, which is resolved in a twofold acceptance both of the *fait accompli* and of the idea of law as a tool serving it. It is hence significant to eschew the belief that legal scholars (in particular, scholars of constitutional law) must simply offer solutions to any request or pressure originating from a phantom and separate effectual reality.

Unlike many of its sociological articulations and concrete developments, legal realism—as I see it above all from a constitutional perspective—values the autonomy of law, because only by preserving this autonomy can scholars mould principles suitable for shaping institutions crucial for people's lives, while at the same time maintaining the viewpoint proper to law.[16]

In this sense, looking at the smart city from the constitutional perspective of equality and social peripheries means looking to law through cities, in order to acknowledge what image and understanding of law is given to us. In my opinion, it is not law understood as a mere technique and therefore adjustable to any purpose, but law bearing an autonomous point of view, because it is propped up by constitutional principles and by fundamental rights.

In fact, the "optimism of reason" postulates a great responsibility of scholars, without which those who enthusiastically and acritically embrace the smart city and its technologies not only abdicate their role but end up being naïve. For this reason, the smart city can have a prescriptive and dulling effect on people, or an effect of openness and coordination[17]: travelling one road or the other is in no way deterministic or neutral but is the consequence of a political and theoretical choice, meaning a choice about what kind of law we require over the smart city.

Thus, after having set aside the idea that smart technologies and agendas are neutral, we must start from the idea that shaping smart cities underlies a preliminary choice relating to the type of interactions between individuals and technology. Law aims to shape these interactions in accordance with lines that, not being rigidly predetermined by the Constitution, are, in spite of this, delimited by it and oriented towards the central importance of human beings. From the methodological standpoint, it is not law that must adapt to the needs arisen with the development of smart cities, by planning solutions necessary to meet these needs,[18] but it is the notion of the smart city that must be brought back within the scope of urban planning, which will be "smart" to the extent that it is focused on constitutional principles.

### 2.1 The Constitutional Understanding of the Connections Between Smart Cities and Social Peripheries, Beyond Artificial Intelligence

The narrative of the smart city is based on some premises whose validity and desirability are often presented as self-evident. Technology, ecologism, sustainability, and inclusion are concepts that go hand in hand with smart cities, but without their being examined from the standpoint of contents nor from that of the means and the not

[16]The relevance of maintaining the autonomous perspective of law is emphasised by S. NICCOLAI, *Principi del diritto principi della convivenza. Uno studio sulle regulae iuris*, Napoli, Editoriale scientifica, 2022, *passim*.

[17]As emphasised by R. SENNETT, *Building and Dwelling*, New York, Farrar, Straus & Giroux, 2018, 158 ff.

[18]For a critique of this acritical and functionalist legal method, see G. AZZARITI, *Diritto o barbarie. Il costituzionalismo moderno al bivio*, Bari-Roma, Laterza, 2021, 166 ff.

merely economic costs for reaching these objectives. So that one ends up making the very error that Simone Weil warned against, according to which "[w]e too easily accept material progress as a gift from heaven, as something that goes without saying; we have to look, face on, at the conditions at the price of which it takes place".[19]

In linking the smart city to technology, one neglects, for example, to consider that technology might not in fact—or at least not necessarily—be an instrument of inclusion and participation, but a factor of socio-spatial discrimination: it has a direct impact on the distribution of resources and opportunities for those who lack the required technological skills. In this sense, technology and financialisation are two factors that can determine a new fracture within the city: no longer—or better, not so much—a *caesura* between a working centre and a periphery lacking infrastructures and services, but a distance (and not merely a spatial one) between individualsn. It is the distance between those who possess the cognitive skills and economic means to benefit from the smart services of information and mobility, and those who, on the contrary, can only take refuge in the narrow and restricted portion of city to which they are relegated.

In essence, the smart city of IT services corresponds to a different socio-spatial division of work and a corresponding new form of variable polarisation of urban space. Within this space it cannot be stated *a priori* that technology and the indistinct space of the network bring with them a fairer distribution of the power of places.

Similarly, doubts may also be raised around other key words in the smart city narratives, like sustainability and ecologism: many goals of smart urbanism are held to correspond to these words, but without their being investigated in their concrete meanings and implications, not just legal but even social and ethical.[20] Sharing these goals may in fact be predicated only in the abstract, because it is required above all to reach understanding on concepts that are not at all self-evident[21] and why they are so easily coupled with urban smartness.

In this regard, in the intent to go beyond the acritical association of ideas between *smart city* and artificial intelligence, which is anodyne in and of itself, and to promote a constitutionalist understanding of the smart city, it may be useful to link this notion to that of urban periphery. This heuristic link makes it possible to understand not only how the smart city relates to the peripheries, but also how and in what sense it can be a vehicle of new forms of periphery and of peripherality.

To this end, it is crucial to assume a multidimensional notion of periphery and peripherality: even from a constitutional perspective this allows us not to limit ourselves to an analysis downstream, which simply acknowledges the inequalities produced or conveyed by smart cities, but to point out, upstream, the point of view that legal scholars and law have of these inequalities.

Centre, periphery, and semi-periphery have constitutional bearing connected directly to the principle of equality and to the comparison that the latter requires in order to establish when there is inequality.[22] The linkage between the concept of periphery and the principle of equality

[19] S. Weil, *Réflexions sur les causes de la liberté et de l'oppression sociale*, It. transl. *Riflessioni sulle cause della libertà e dell'oppressione sociale*, Milano, Adelphi, 1992, 70–71.

[20] Cf. M. Ziosi, B. Hewitt, P. Juneja, M. Taddeo, L. Floridi, *Smart Cities: Reviewing the Debate About their Ethical Implications*, supra note 5.

[21] For a critical analysis of the notion of sustainability and of its function of quieting dissent and conflict, see the contributions contained in R. Krueger, D. Gibbs (eds), *The sustainable development paradox. Urban political economy in the United States and Europe*, New York-London, The Guilford Press, 2007; see also E. Swyngedouw, M. Kaika, *Re-naturing cities: great promises, deadlock . . . and new beginnings?*, in K. Archer, K. Bezdecny (eds), *Handbook of Cities and the Environment*, Cheltenham-Northampton, Elgar, 2016, 50 ff.

[22] We are referring to the fact that, given two cases or situations, from the legal standpoint they may be deemed (and be treated as) equal or unequal not in and of themselves, because they will always have shared elements and elements of diversity (otherwise they would be identical). They may be deemed equal or unequal depending on whether shared or different elements are assumed, and therefore always in relation to something else, which functions as an element of comparison.

descends from the relative and relational scope of the concept of periphery and peripherality: there is no centre, periphery or semi-periphery that is such indefinitely, or that is so in and of itself; there is a centre, periphery, or semi-periphery in relation to a given space-time, to an object of study and in relation to magnitudes and parameters that are not exclusively geographic in nature.

It is important, then, to remark that neither cities nor peripheries are geographical places alone. First and foremost one cannot think of peripheries merely on the basis of the physical distance from a centre, nor just because they mark the boundary between urban space and rural areas, but to the extent that they are conceived and experienced in terms of settlement functional to city' logic of production, growth, and exclusion.

This is illustrated, for example, by the fact that, from the socioeconomic and cultural standpoint, the peripheral areas of many European cities have features appreciably different from, for instance, certain North American suburbs.

In Europe the term "periphery" is used with a negative connotation, to allude to the ghettoization of the people living in peripheral neighbourhoods and to the shortcomings of the services situated in these areas. In North America the peripheral areas of cities are often marked by gated communities, consisting of high-level dwellings for middle and upper-class families intentionally separated from the surrounding environment.[23]

Therefore, just as the term "periphery" has in Europe become mostly synonymous with poverty and social marginalisation, elsewhere (particularly the United States[24]) the peripheral urban space symbolise distinction and self-isolation, in accordance with those "geographies of centrality that connect the power centres of the world and cut across the old North–South divide".[25]

Alongside—or rather, in addition to—the spatially defined peripheries, there are in fact socially characterised peripheries, namely areas of the city that, regardless of location, show signs of urban decay and of scant social integration. In this sense, the conditions of social hardship and of economic isolation are not an exclusive prerogative of the more decentralised areas, but are also found in relatively central neighbourhoods, which may therefore be defined as social periphery.

In this additional meaning, then, urban peripherality does not result from a geographical position at the margins of the metropolitan area, but derives from more complex and often interwoven elements, such as: low quality of or limited access to urban infrastructure services (transport, water, waste, energy, and communications); low-quality or deteriorated buildings and infrastructures; non-proximity to quality educational and cultural institutions, public spaces, knowledge infrastructures, hospitals, and other municipal structures; limited or absent spread of knowledge, culture, skills, civic participation, social equity, and social innovation; concentration of people excluded

[23] Thus G. AMATO, *Democrazia e redistribuzione. Un sondaggio nel Welfare statunitense*, Bologna, il Mulino, 1983, 14.

These are, then, forms of urban settlement that, although less common, are spreading in Europe as well: see P. MARCUSE, R. VAN KEMPEN, *Conclusion: A Changed Spatial Order*, in P. MARCUSE, R. VAN KEMPEN (eds), *Globalising Cities. A New Spatial Order?*, Malden-Oxford-Carlton, Blackwell, 2000, 254 f.; in a legal perspective, see L. RON, *Gated Communities in Law's Gaze: Material Forms and the Production of a Social Body in Legal Adjudication*, in *Law and Social Inquiry*, 2009, 635 ff.

[24] In any event, there are more than a few signs of a countertrend: cf. C. BARRINGTON-LEIGH, A. MILLARD-BALL, *A century of sprawl in the United States*, in *Proc Natl Acad Sci USA*, 27/2015, 8244.

[25] S. SASSEN, *Urban Gating: One Instance of a Larger Development?*, in S. BAGAEEN, O. UDUKU (eds), *Gated Communities. Social Sustainability in Contemporary and Historical Gated Developments*, London-Washington, Earthscan, 2010, XI. But see also S. SASSEN, *Cities in a World Economy*, Thousand Oaks, Pine Forge Press, 2000, 78, who, in particular regarding Europe, emphasises the presence of a multiplicity of geographies of centrality and of multiplicity: "Cities at the periphery will feel the widening gap with the newly defined and positioned geography of centrality" (*ibid*, 45, my italics).

With reference to the European dynamics and to the hierarchies between centre and peripheries, cf. D. KUKOVEC, *Law and the Periphery*, in *European Law Journal*, 3/2015, 406 ff.

from economic activity; concentration of socially vulnerable people (for example low-income households, the elderly, immigrant communities) stricken by urban marginalisation (high crime rates, low graduation rates, etc.).[26]

This helps to reveal two hurdles: *i.* the discriminatory outcomes that the spatial location of things, people, and relationships has on the actual guarantee of constitutional rights; *ii.* the way in which law and legal scholars contribute towards creating or consolidating spaces of marginalisation and social exclusion, also through the acritical use of the smart city concept.

For this reason, public intervention in peripheries and for peripheries demands awareness of a double complexity: *i.* the peripheral facade of an area does not result merely from its spatial location; *ii.* an area can be deemed peripheral the more it adds up peripheralities related to the different activities (social, cultural, commercial, etc.) that define the quality of life.

## 3 Smartness, Social Peripheries, and Inequalities: A New Opportunity for Multiple Peripheralities?

According to an optimistic narrative of the smart city,

> [t]he incorporation of suburban and marginal districts into strategic planning for smart-city agendas is an encouraging trend: to address the social and spatial marginality of peripheral urban areas and enhance their resilience, it is necessary to overcome entrenched conceptual and material forms of core-periphery opposition. Initially, this tends to occur with peripheral areas being identified as 'integrated smart regeneration areas' – that is, areas targeted for systematic transformation via the convergence of building redevelopment interventions, cultural heritage and landscape designation, economic revitalisation, infrastructure investment, and the strengthening of citizenship services through innovative technologies.[27]

Peripheries, then, would have everything to gain from incorporation into a smartness-oriented urban planning. Relying on smart regeneration and urban requalification, smart planning would provide strong impetus towards overcoming the opposition between centre and periphery, increasing what would appear to be the inherent resilience of peripheries.

However, this reading is also founded upon a set of undemonstrated assumptions.

It is needed in the first place to examine the connections between smart city, the capitalism of austerity, and the resilience of the peripheries, to understand the extent to which the rhetoric of resilience is deceptive pandering contradictory to the constitutional perspective of substantial equality. The resilience acritically associated with peripheries appears actually to be the result of so-called austerity politics, which postulate the silent acquiescence to a creeping pressure on social rights above all, but also on civil rights.[28]

On the other hand, as far as the insistence upon urban regeneration is concerned, the costs –not only the financial ones—of an urban regeneration devoted to or guided by smartness must be questioned.

In fact, if one considers urban requalification programmes and urban regeneration projects, aimed also at reducing the phenomenon of marginalisation and social decay, ambivalent conclusions are reached: the partnership between public institutions and private investments upon which they rest sometimes ends up impacting the choices of public actors, so that it is the private interests that guide public sectors, as the general

[26] See S. de Falco, M. Angelidou, J.-P. D Addie, *From the "smart city" to the "smart metropolis"? Building resilience in the urban periphery*, in *European Urban and Regional Studies*, 2/2019, 209.

[27] S. de Falco, M. Angelidou, J.-P. D Addie, *From the "smart city" to the "smart metropolis"? Building resilience in the urban periphery*, 219, supra note 26.

[28] On the misuse of the term "resilience" and its constitutional repercussions, see N. Pettinari, *La positivizzazione normativa della 'retorica della resilienza'. Riflessioni sulla crisi dei fondamenti pluralisti e sociali nel costituzionalismo italiano del tempo presente*, in *Nomos*, 2/2022, 9 ff.

interest retreats. This is demonstrated by the regenerations of peripheral neighbourhoods, which were transformed into the driver for new forms of gentrification (such as austerity or green gentrification[29]) and therefore of displacement and expulsion from cities. Local administrations, in fact, are not always able to guide the urban transformation initiatives of private players towards social objectives.

In similar cases, the renovation of peripheries can result in the creation of new peripheries and social marginalisation as it does not act upon the causes of urban inequalities. This is because urban regeneration (even if smart) cannot be a solution for the imbalances between centre and peripheries when it is not conceived and implemented in discontinuity with the urban development models adopted thus far.

Ultimately, the smart city of IT networks, technologies, and digital services points out a different socio-spatial division of labour and a new form of variable polarisation of the urban space, but the indistinct place of networks does not bring, in and of itself, a fairer distribution of power nor a worthwhile enhancement of peripheral areas. Although information technologies disseminate production and exchanges, they do not cancel the relevance of the spatial factor. On the contrary, they contribute towards concentrating, within global cities, services and control over geographically dispersed economic operations, with the creation of new geographies of centrality and marginalisation.[30]

This is also recognized by those who initially gave an optimistic reading of the so-called "creative class" of smart cities, to which they associated positive effects of urban and social improvement. A few years later, they actually recognized that the rise of the creative class has had highly contradictory effects on cities: the inequalities between large, globalised cities and smaller urban centres grew greater, cities came up against a strong gentrification process, and the gap between centres and peripheries widened.[31]

This point raises inevitable legal questions, as law should not simply translate the ambition of creating smart cities into regulations, just as such regulations should not be adopted in the interest of profit and profitable investments. Otherwise, in the absence of public guidance and notwithstanding declared intents, infrastructures will tend to be placed where there are greater expectations of economic return, with paradoxical effects of fragmentation of the places and displacement of those who inhabit them.[32]

## 4 What Constitutional Point of View About Smart City and Social Peripheries? Raising Questions Before Giving Answers

It follows from the foregoing considerations that a smart city is not necessarily a just city,[33] because it is not necessarily one that is fairer or more sustainable. It is therefore important to reject that devastating form of objective alienation that is the reification of the smart city, and to start from the idea that the smart city is a container.

Before thinking of smart solutions for cities, it is necessary, then, to discuss the typology of smart city that is to be built, so it can be inhabited by embodied people, who refer to things as goods before commodities and in it entertain meaningful human relationships and not only economic

[29]See for example S. ANNUNZIATA, L. LEES, *Resisting 'Austerity Gentrification' and Displacement in Southern Europe*, in *Sociological Research Online*, 21(3)/2016, 148 ff.; F.M. ALEXANDRESCU, L. PIZZOL, A. CRITTO, *Green gentrification as strategic action: Exploring the emerging discursive and social support for the Green Tree Strategy in Porto Marghera, Italy*, in *Cities*, 118/2021, 1 ff.

[30]S. SASSEN, *Cities in a World Economy*, London, Sage, 2019, 319 f.

[31]R. FLORIDA, *The New Urban Crisis: How Our Cities Are Increasing Inequality, Deepening Segregation, and Failing the Middle Class and What We Can Do About It*, New York, Basic Books, 2017.

[32]On this point, see S. GRAHAM, S. MARVIN, *Splintering Urbanism. Networked Infrastructures, Technological Mobilities and the Urban Condition*, New York, London-New York, Routledge, 2001.

[33]Cf. S.S. FAINSTEIN, *The Just City*, Ithaca-London, Cornell University Press, 2010.

transactions.[34] For this purpose, those who look to the smart city from the perspective of constitutional law and recognise law's fundamental autonomy (which does not mean detachment) are tasked with problematizing this notion in light of the Constitution.

In this sense, the Constitution provides the tools to recognize inequalities in all their countless guises, even smart ones, providing that people located in these places, relationships and experiences are not lost sight of.

The constitutional vision must be enhanced, as it takes priority over the mere *ex post* regulation of the smart city. This regulation, in fact, very often follows or runs after the self-regulation of private actors, adjusting to the privatisations of public spaces in the interest and for the sole profit of investors.

From the constitutional standpoint, a public intervention is not sufficient if it goes no further than the mere control of "big tech" and of digitalised capitalism, in accordance with the idea of the market that limits and regulates itself. It is required, rather, to keep the regulation of the smart city from being paired with the so called "nudge regulation"[35] and its underlying libertarian paternalism, so that the realisation of the smart city might not be instrumental and subservient to a sort of biopolitical urbanism.

[34] See A. BELLAN, *La Scuola di Francoforte e il problema della reificazione*, in *Costruzioni psicoanalitiche*, 2012, 101 f.

[35] R.H. THALER, C.R. SUNSTEIN, *Nudge: Improving Decisions About Health, Wealth, and Happiness*, New Haven, Yale University Press, 2008.

# Smart Working, Its Places and Some Forward-Looking Directions

Enrico Gragnoli

**Abstract**

The pandemic brought about an unexpected technological turning point and faced all elements of the labour market with a common question: is it necessary or even simply convenient to transform the way of working, with the partial or total liberation of workers from the obligation to be present? Smart working has opened up new scenarios and leads to important insights in several respects: given this, the paper analyses the success of individual agreements in business organisations, the places of smart working, the employee's interest and the determination of pay, suggesting some innovative perspectives for the future.

## 1 Smart Working, Subordination in Employment and the Epidemics

Not only has smart working called the entire category of employment subordination into question and raised theoretical problems, which defy any easy solution and which, in all truth, have not yet been fully set,[1] not least due to the very little application experience. However, it has also clearly shown, if proof were needed, significant social innovation and has blurred the boundaries between time for personal life and time for occupational activity.[2] This resulted in the specification of the right to log out,[3] meaning getting away from all electronic connections to safeguard time for intimacy and time exclusively dedicated to family and personal relationships, with no risk of any oppressive trespassing of information and requests associated with the interests of the company or of the public administration.

The disruption of businesses and institutions as places of work appears to have already been inescapable in Italy since the end of 2019, regardless of the onset of the pandemic, which, in the space of a few weeks, raised brutal questions about matters of organisation and the protection of public health and safety. Indeed, precisely during those years, the difference between white-collar and blue-collar jobs became wider, as the former can be partly done in places chosen by the

[1] See: Caruso, *Il lavoro digitale e tramite piattaforma: profili giuridici e di relazioni industriali. I lavoratori digitali nella prospettiva del Pilastro sociale europeo: tutele rimediali, giurisprudenziali e contrattuali*, in *Dir. rel. ind.*, 2019, 1005 *et seq.*

[2] See: Signorini, *Il diritto del lavoro nell'economia digitale*, Torino, 2018, 15 *et seq.*

[3] See: Spinelli, *Tecnologie digitali e lavoro agile*, Bari, 2018, 137 *et seq.*

E. Gragnoli (✉)
Modena, Italy
e-mail: enrico.gragnoli@studiogragnoli.eu



A. D'Aloia, S. Scagliarini (eds.), *Law and Rights in the Smart City: The Challenge of Regulation*, The City Project 9, https://doi.org/10.1007/978-3-032-13071-6_3

worker, whereas the latter must be done in predetermined places, with strict working hours and often perceived as degraded, also for this very reason. For intellectual work, the physical break-up of the firm went alongside the progressive fading of workers' collective cohesion and of the force of their trade union affiliation, even though the related associations try to obviate this through direct electronic communication,[4] strengthening contact between members and their representatives.[5]

Although with no epidemic, smart working and its success would leave our workers freer (or, at least, convinced of being freer), but certainly more alone, deprived of the human experience associated with their occupational commitment at the company and immersed in telecommunication connections, which, at least to fossils like I am, seem a little pitiful, in a flood of e-mails, data, summary news, which make it difficult to understand their thread and sense. If these considerations, which cannot be proved and go against their favourable reception by scholars, are accepted,[6] smart working would have nonetheless ended up reducing cohesion, culture and the personal ability to interpret the action of the company and of its management, while facilitating phenomena of isolation and subjection to the standpoint imposed by the employer, to the detriment of freedom, at least within the company.

When many activities considered non-essential were forbidden, as of 7 March 2020, this was accompanied by the constant and somewhat distasteful call for smart working, which was misunderstood by the political system and confused with the simple performance of work in the absence of physical contact and, as far as possible, carried out from home, in complete isolation.[7] As often happens in the Italian tradition, the response given by employers and workers proved less disastrous than the institutional one and, in line with the ability to improvise that has characterised our people since time immemorial, with no planning whatsoever, our administrations, companies and self-employed workers survived, albeit with a drastic reduction in the gross domestic product.

In all the cases in which it proved feasible (and there were many), work was performed through electronic resources,[8] which were available before but were, for the most part, used for the first time in such an intensive and widespread manner, because of a state of necessity. Companies and institutions realised that they could deploy tools that ensured and can permanently ensure their survival with the employees working from home.[9] For companies, it remains to be seen whether these methods can ensure reasonable profitability, in other words, ensure not just the continuity, but the efficiency of their traditional functions, above all without intruding too much into workers' private lives[10] and into their[11] work-life balance.[12] In any case, however dramatic it may have been, the pandemic brought about an unexpected technological turning point and faced all elements of the labour market with a common question: is it necessary or even simply convenient to transform the way of working,

[4]See.: Marazza, Social, *relazioni industriali e (nuovi percorsi di) formazione della volontà collettiva,* in *Riv. it. dir. lav.*, 2019, I, 57 *et seq.*

[5]See: Occhino, *Nuove soggettività e nuove rappresentanze del lavoro nell'economia digitale,* in *Labor,* 2019, 5 *et seq.*

[6]See: Timellini, *In che modo oggi il lavoro è* smart? *Sulla definizione di lavora agile?,* in *Lav. giur.*, 2018, 230 *et seq.*

[7]See: M. Miscione, *Il diritto del lavoro ai tempi orribili del* coronavirus, in *Lav. giur.*, 2020, 321 *et seq.*

[8]See: Filì, *Diritto del lavoro e dell'emergenza epidemiologica da* Covid-19 *e nuova "questione sociale"*, in *Lav. giur.*, 2020, 332 *et seq.*

[9]See: D. Garofalo, *La dottrina giuslavorista alla prova del* Covid-19: *la nuova questione sociale,* in *Lav. giur.*, 2020, 429 *et seq.*

[10]See: Tinti, *Il lavoro agile e gli equivoci della conciliazione virtuale,* in *WPSDCLE "Massimo D'Antona"*, it., no. 420, 2020, 7 *et seq.*

[11]See: Malzani, *Il lavoro tra opportunità e rischi per il lavoratore,* in *Dir. merc. lav.*, 2018, 17 *et seq.*

[12]See: R. Pessi – Fabozzi, *Gli obblighi del datore di lavoro in materia di salute e sicurezza,* in Aa. Vv., *Il* Jobs act *del lavoro autonomo e del lavoro agile,* edited by Perulli – Fiorillo, Turin, 2018, 227 *et seq.*

with the partial or total liberation of workers from the obligation to be present[13]?

In the social scope, the questions are consistent[14] and were posed several years ago,[15] so that their weight and urgency have simply increased. This entails a qualitative change, as, while the evolution would have been gradual and led by progressive experimentation,[16] the events of 2020 brought about rapid changes, with improvisation and not much consideration. Italian Law no. 81 of 2017 had been designed precisely for a process of possible (and not even certain) modernisation[17] of public administrations and enterprises[18] and was overshadowed by the urgency of contagion prevention.

## 2 Smart Working and the Success of Individual Agreements in Business Organisations

Irrespective of the epidemic, smart working gives further evidence of a more general trend in this century, featuring progressively wider use of individual agreements, first as regards the determination of pay, and then organisation, with the consensual identification of the methods and expected results, then the criteria for their identification, albeit in fully or partly remote cooperation. At large manufacturing companies having higher technology and automation, at structures engaged in sales and contact with customers and vendors and in management, executive and coordination areas, the number of workers who are appreciated for their quality and who can obtain commensurate remuneration is increasing.

This is quite frequent in the manufacturing industry, as, because of the high complexity of capital goods, individual contributions have become more important, and skills have effects on profits or loss. In this scenario, which is the result of the use of sophisticated IT and telecommunication technologies, remuneration is no longer determined through collective bargaining and instead, individual bargaining has acquired a significant or even prevailing role.[19] A massive change in resources occurred and their complexity made their performances no longer fungible. Only skilled people can give a contribution in line with expectations and with the requirements of frantic competition.

As pay is determined in individual agreements,[20] many are able to choose the employer that gives them the highest remuneration for their skills (skill-based pay),[21] albeit with risks associated with age, personal difficulties or physical conditions. At the highest levels of knowledge, there is now competition between workers in seeking the most stimulating opportunities, which generates the risk of fast obsolescence of knowledge of those who are not as quick in staying up to date or unable to adapt to the endless changes. Due also to the fact that many functions are being progressively moved to remote countries, which have less effective social protection and income safeguard systems than Italy, pay is now at the centre of increasing tension between the classic categories of labour law and

[13] Before the epidemic and irrespective of its massive impact, see Verzaro, *Fattispecie della prestazione agile e limite dell'autonomia individuale,* in *Riv. it. dir. lav.*, 2019, I, 254 *et seq.*

[14] See: Treu, *Rimedi, tutele e fattispecie, riflessioni a partire dai lavori della* gig economy, in *Lav. dir.*, 2017, 393 *et seq.*

[15] See: Proia, *L'accordo individuale e le modalità di esecuzione e di cessazione della prestazione di lavoro agile,* in Aa. Vv., *Il* Jobs act *del lavoro autonomo e del lavoro agile,* edited by Perulli – Fiorillo, Torino, 2018, 179 *et seq.*

[16] See: Donnini, *Nuova flessibilità spazio temporale e tecnologie: l'idea di lavoro agile,* in Aa. Vv., Web *e lavoro. Profili evolutivi e di tutela,* Turin, 2017, 87 *et seq.*

[17] See: Tiraboschi, *Il lavoro agile tra legge e contrattazione collettiva: la tortuosa via italiana verso la modernizzazione del diritto del lavoro,* in *Dir. rel. ind.*, 2017, 962 *et seq.*

[18] See: Franza, *Lavoro agile: profili sistematici e disciplina del recesso,* in *Variaz. temi dir. lav.*, 2018, 773 *et seq.*

[19] See: Mazzotta, *Autonomia individuale e sistema del diritto del lavoro,* in *Giorn. dir. lav. rel. ind.*, 1991, 489 *et seq.*

[20] See: D'Antona, *Contrattazione collettiva e autonomia individuale nei rapporti di lavoro atipici,* in *Giorn. dir. lav. rel. ind.*, 1990, 540 *et seq.*

[21] See: Dell'Olio, *Retribuzione, quantità e qualità di lavoro, qualità di vita,* in *Arg. dir. lav.*, 1995, 1 *et seq.*

those of a new model,[22] which replaces collective bargaining agreements with individual ones and equal pay with skill-based pay,[23] generating the risk of skill ageing and of the impossibility to find a job in line with aspirations. The existential and occupational condition of those aspiring to bargain in order to have merit rewarded is different from that of those who rely on collective bargaining agreements and, therefore, on a solidarity dimension in order to obtain a solidarity-based protection of their weakness.[24] Despite the heated battles for their renewal, national or company collective bargaining agreements are irrelevant to those who receive pay that is far higher than the minimum; indeed, pay increases do not change the amount received, in accordance with the "*principio dell'assorbimento*" ("absorption principle").[25]

The advent of smart working has shifted the relevance of individual agreements towards an aspect of corporate dynamics that was little considered before, i.e. towards the organisation and specification of activity, with bargaining set between the company and the individual worker, seeking personal balance, along with all related pitfalls, including poor ability to plan. Smart working is not so much a tool reconciling personal and occupational requirements as the shifting from a collective dimension to an individual one in the corporate setting, with a different meaning of subordinated employment and, most of all, with bargaining on how jobs are performed.

In line with the unrelenting success of electronic developments, the process of every company and of many institutions has changed, with a higher number and importance of functions that can be performed anywhere and, thus, with dematerialisation not so much of cooperation, but of its underlying assumptions, following a trend that was already firmly established and has recently become even stronger.[26] For those who love tradition and still hold the twentieth century concept of work (because of their age for example, or unfamiliarity with the latest resources), these changes mark the risk of a solipsistic attitude, of progressive isolation from fruitful human interaction and of a productivity-oriented view of activity, aimed at immediate measurable results, with estrangement not only from long-lasting and dialectic relationships with colleagues and superiors, but also from the very work environment, which may be oppressive, but (hopefully) also the source of opportunities for critical thinking.[27]

## 3 The Places of Smart Working

Not only does smart working mark the transformation of companies and institutions, which are now ready to welcome the new, but also the readiness of employees, for various reasons.[28] On the one hand, electronic resources have proved fascinating, as workers believe they have more freedom working out of the office for at least some days; on the other hand, the use of electronic devices has become familiar, and they are centred on performing tasks and partially indifferent to the place where the activity is performed. The response of employers has been varied but not such as to discourage smart working, with an application of the rules not even comparable with that of the stricter telecommuting schemes, which went live in the late 90 s and have never achieved appreciable results, other than in public administrations, where they are governed by exhaustive regulations,[29] which are conversely lacking in the

[22] See: Liso, *Autonomia collettiva e occupazione,* in *Giorn. dir. lav. rel. ind.,* 1998, 220 *et seq.*

[23] See: L. Zoppoli, *La corrispettività nel contratto di lavoro,* Napoli, 1991, 157 *et seq.*

[24] See: Magnani, *Il salario minimo legale,* in *Riv. it. dir. lav.,* 2010, I, 769 *et seq.*

[25] See Italian Court of Cassation, ruling no, 14689 of 29 August 2021, in *Giur. it. rep.,* 2012.

[26] See: Occhino, *Il lavoro e i suoi luoghi,* Milano, 2018, 125 *et seq.*

[27] See: Corso, *Sfide e prospettive della rivoluzione digitale, lo* smart working, in *Dir. rel. ind.,* 2017, 979 *et seq.*

[28] See: Casillo, *La subordinazione "agile",* in *Dir. merc. lav.,* 2017, 545 *et seq.*

[29] See: M. Miscione, *Diverse tipologie contrattuali:* remote working, *telelavoro* e digital workplace, in *Lav. giur.,* 2009, 663 *et seq.*

private sector. Rather, Italian law no. 81 of 2017 caught attention for the activity results.[30] Although there was actually no need for it, this process has incentivised employees' readiness to use electronic solutions, with the lure of escaping (at least for part of the time) from the company, which is considered oppressive or restricting on life opportunities. On the other hand, with growing acceptance of control based on merit and, therefore, on the achievement of assigned productivity objectives[31] (a condition for smart working), contrary to the twentieth century tradition, the image of the place of work as a social context promising the establishment of long-lasting relationships has been fading.

The home is becoming more and more a place for work and this requires some considerations about its design, as its structure and spaces must be consistent with more complex functions. Indeed, at least in Italy, home offices, designed to be used by people working remotely for firms located elsewhere, have only modest success. Conversely, there is quite a favourable reception of being posted away from the company headquarters, to other provinces, other countries, to places where work can be combined with leisure, granted that said combination is successful and does not make either of the goals prevail fully over the other. Nonetheless, for remote working, for white-collar jobs, work is increasingly shifting to the home, with the unavoidable sacrifice of cohabitation at the company, which has always been regarded as one of the core features of work under the direction of others.

Given that, after the acute epidemic phase, smart working is a free choice, at least in companies, employees can assess whether this method of performing their duties aligns with their interests, thus being ready to open their private places to the intrusion of different activities; their combination may prove less than easy, as, on the one hand, homes were mostly designed in a different regulatory scenario and, on the other hand, in terms of functionality, similar interconnections between family and productive experiences may generate problems, evidence of which is given by the focus on disconnecting. In architectural and town-planning terms, the establishment of a material percentage of remote work activities requires that the very purpose of homes be considered, along with the needs of those living in them, with most properties having been conceived in different ages and a clear divide between the expectations of those aspiring to or rejecting remote work.

In Italian Law no. 81 of 2017, the methods to perform work duties shall be determined with individual agreements, which shall establish how the job can be performed without the employee being present and how the employee's activity can be coordinated with the remaining corporate or institutional functions[32]; after all, for actual control over results and for the freedom granted, this solution is unavoidable, as only the agreement between the employer and the worker can define how their cooperation shall be performed and, if the latter enforces his or her personal reasons, the former must consider its production requirements and the possibility of meeting them.[33] After all, a point of balance must be reached between aspirations to the best allocation of one's time and to reconciling personal and work commitments and, on the other hand, the preservation of profitable results and, where possible, better ones that those achieved with a more traditional cooperation approach.[34]

---

[30] See: Andreoni, *Il lavoro agile nel collegamento negoziale,* in *Riv. giur. lav.,* 2018, I, 106 *et seq.*

[31] See: Martone, *Lo* smart working *nell'ordinamento italiano,* in Aa. Vv., *Giuseppe Santoro Passarelli, giurista della contemporaneità,* Liber amicorum, Turin, 2018, vol. II, 1026 *et seq.*

[32] See: Proia, *L'accordo individuale e le modalità di esecuzione e di cessazione della prestazione di lavoro agile, loc. cit.,* 179 *et seq.*

[33] See: Pinto, *La flessibilità funzionale e i poteri del datore di lavoro,* in *Riv. giur. lav.,* 2016, I, 367 *et seq.*

[34] See: Tiraboschi, *Il lavoro agile tra legge e contrattazione collettiva: la tortuosa via italiana verso la modernizzazione del diritto del lavoro, loc. cit.,* 965 *et seq.*

## 4 Smart Working and Employees' Interest

It is generally believed, with precise evidence given by frequent behaviours, that smart working is a solution favourable to the worker, who often asks for it, at least in order to avoid the risks, costs and delays associated with traffic and public transport. Somehow, home as a shelter is preferred over corporate life communion, and isolation has reacquired the fascination that seemed to have been forgotten from the last century, in the name of the relationships being facilitated by work contacts. However, since the end of the epidemic, signing the related agreement is a free choice to be made by the company and, consequently, it must have some material interest in it, which is seldom the best possible allocation of spaces, rational arrangement of offices and reduction of the related expenses. The not always easy negotiation between the employer and the worker mostly develops around the pursuit of greater efficiency, hoped for by the former and promised by the latter, and therefore the subject-matter of bargaining and of the agreement on the job to be performed electronically or in any case remotely.

Referring a large part of organisational matters to individual agreements requires discussion on how opposite needs can be met, and the employee's wishes clash with those of the employer, outside the scope of the more reassuring collective bargaining agreements and, however, as an alternative to the workplace dynamics and the related community of life. Seeking their own convenience (real or imaginary), workers must convince the company (and, with lesser difficulty, the public administration) and they must do so by convincing them that their contributions would improve if placed in a different organisational model. Therefore, in view of a mutual advantage, but with the company in the stronger position, efficiency becomes a topic for discussion and, in that, it changes subordination, due to said consensual and dialogue components assumed by the exercise of direction power, which yields to the agreement about the place and way that work is performed.

Fascinated by the idea (perhaps bordering on delusion) of recovering room for freedom and wrongly believing that everything can be reconciled thanks to technology, as if it ensured dominion of time, smart workers do not realise that they must give a lot, especially if the company and the public administration are able to measure their contributions precisely. If the intention is to … escape from the company, not only must the contribution be unchanged, but it must also meet the company's expectations to such an extent as to justify the hoped-for consent. The points of balance of these endless negotiations defy any satisfying classification and this is unavoidable exactly because they belong to an individual dimension.

The related breakdown of the corporate structure diminishes the community dimension of work and entails drifting towards individualism and often hectic paces. Being isolated at home or at some other place that they choose, employees (especially private sector employees) must justify their choice proving the advantage for the employer and, therefore, gaining accreditation as the providers of distinctive contributions, which are enabled or facilitated by the granting of their requests.

On the other hand, smart working sometimes is grounds for ever latent anarchist sentiments to arise, with the poorly hidden aspiration of many middle or high ranking employees to set themselves free from the directions given by the senior management and to acquire material decision-making autonomy. If anything, in that scenario, reaching individual agreements proves more difficult, as they must ensure that the impartial objectives set by public administrations be pursued, achieving public interest.[35] This assumes controls that are objective or, in any case, less discretionary than those exercised by companies, because the trust logic informs civil service to a lesser extent than work in the private sector, and the former features stronger and more stable

[35] See: Zilli, *Il lavoro agile nella pubblica amministrazione 4.0,* in Aa. Vv., *La nuova frontiera del lavoro: autonomo, agile, occasionale,* edited by D. Garofalo, Adapt university press, Bergamo, 2018, 356 *et seq.*

requirements for employees' behaviours to be compliant, at least in order to ensure equal treatment.

## 5 Remote Work and Determination of Pay

For certain professional vocations, it has been possible for some time now to carry out work abroad, and the epidemic has caused this issue to arise to an unexpected degree. With no specific difficulties, employees posted to different countries can work under subordinate employment schemes, to the full satisfaction of companies, which sometimes do not even know and do not really care to know where their workers are and whether they live in the same country where the company is based. Partly because of the pandemic, cross-border smart working has become a significant phenomenon.

For the time being, there are no realistic prospects of European Union-wide trade union agreements and no indications from the European Union. The resolution adopted by the European Parliament on 21 January 2021 concerns the different profile of disconnection and acknowledges the importance of remote work, but says nothing about remuneration. Because of the lack of any initiative of the European Union, the parties are free and, specifically, companies operate with structures located in different countries, signing agreements in one country or another based on financial advantages and with work often provided to the group, i.e. to companies other than the one that signed the original agreement. While being typical to multinational organisations, this model is imitated by smaller entities, as they use contributions acquired by other firms and made available to the end recipient, with questionable contracting schemes, as in the case of foreign companies providing phone sales services, recruiting locally and setting up permanent cooperation schemes with Italian companies.

The spread of smart working, always in accordance with the different national frameworks and with the related pay prospects, requires that a decision be made on social interest, as the risks of competition between workers from other countries are growing exponentially and are creating a scenario that, for the time being, is not regulated. Mostly, any definition of the objectives is lacking, whether to allow companies to operate at lower costs, whether to allow employees from less thriving countries to seize opportunities, thereby effectively competing with the others, or whether to protect the latter and their prospects with the companies based in their own country. It is hard to imagine any balance between such different expectations, because they defy any compromise and the European Union will sooner or later have to choose.

In the medium term, its action is unavoidable in the scope of cross-border remote work; indeed, it must determine what interests deserve priority protection, which is no easy task in terms of strategy and of comparison of different expectations. In fact, the opportunity to work across different countries is a reality that, though perhaps inevitable, was accentuated by the pandemic. If we are not mistaken, at the moment there is a rush towards self-employment, whereas many working relationships can be described as employment ones, at least with regard to our legislation.

The applicable collective bargaining agreement had a founding national reach, with the structural limitation of competition and with the implicit prohibition imposed on companies not to compete to the detriment of workers and not to reduce labour cost below the level laid down by mandatory clauses, having regard to Article 36 of the Italian Constitution.[36] The agreement is still national, both because that is its geographical area of application and because it lays down harmonised conditions for all players in the same product sector, across the entire country. Conversely, cross-border remote work connects employees' life requirements rooted in a State and the performance of the activity in favour of a foreign company. How is pay to be determined?

For the time being, at least under Italian legislation, the company decides, applying the same

[36] See Italian Court of Cassation ruling no. 27591 of 14 December 2005, in *Riv. crit. dir. lav.*, 2006, 556.

collective agreement to all workers, irrespective of the methods of cooperation and, therefore, also in the case of smart working. This may allocate additional and unhoped for resources to those living in lower-income European Union Member States, a phenomenon that was significant during the epidemic; the use of electronic resources enables them to enjoy Italian levels of pay in a social context characterised by a lower standard of living. Conversely, if, working from other countries, employees were paid in accordance with the standards in force in their territory, they would bring fierce competition to Italian workers and might jeopardise employment prospects, especially for vocational profiles little affected by language proficiency or familiarity with aspects of the national culture, such as knowledge of law.

Do those working remotely for foreign companies deserve the pay given in the State they live in or in that where the company is based? Adopted in order to regulate competition between employees, collective bargaining agreements, and Italian ones in particular, have never been designed to protect foreigners, and have always had a national reach, at least with regard to determining wages. The Italian scenario is the widest reach of the agreement, granted that it can regulate it with sufficient credibility. Indeed, experiences with supranational agreements have always been quite irrelevant in absolute terms and, all the more so, as regards Italy. It remains to be determined what developments will occur in cross-border remote working.

# Data Protection and Data Sharing in the Smart City Context: A Possible Re-use of Company Data? A Data-Driven Strategy for the Development of Innovative Urban Projects

Carlo Rossi Chauvenet and Sveva Ianese

**Abstract**

The expression "smart city" refers to a polyvalent concept whose main characteristics are still undefined. According to EU Commission, this concept goes beyond the mere application of digital technologies within the urban context and covers a bundle of initiatives devoted to efficiency and social equality. In this perspective, the technological dimension must be added to the social one to obtain a holistic urban renewal. But what does "technological dimension" really mean? And what is the significance of "smartness"? This chapter tries to answer both these questions and then examines the new legal framework for many data sharing and reuse initiatives within the urban context, offering a new perspective for the development of innovative projects.

C. R. Chauvenet
Department of Legal Studies, Bocconi University, Milan, Italy

S. Ianese (✉)
Department of Economics and Management, University of Padova, Padova, Italy
e-mail: s.ianese@crclex.it

## 1 Introduction

The expression "smart city" has been used over time to define different phenomena of civic innovation and redesign.[1] It is a polyvalent concept whose main characteristics are still undefined.[2] In fact, researchers and Public Authorities have provided conflicting definitions of this urban evolution over time.

In particular, according to the European Commission, a smart city can be defined as "a place where traditional networks and services are made more efficient through the use of digital solutions for the benefit of citizens and businesses that inhabit it.".[3]

This conceptualisation considers smart cities as something that goes beyond the mere applica-

[1]This concept has been described as "nebulous" by scholars, as pointed out in Halegoua G. R., *Smart Cities,* Cambridge 2020, 1.

[2]The only reference to the smart city concept seems to be the one provided by Art. 20 of D.l. 179/2021 conv. in L. 221/2012, which refers to the "smart communities" paradigm. See Cartei G. F., *La dimensione smart dello spazio urbano: smart city e rigenerazione urbana.* In: Ferrari G.F., *Smart city (L'evoluzione di un'idea)*, Milano-Udine 2020, 105–115.

[3]EU Commission, *Smart cities*. Available at: https://commission.europa.eu/eu-regional-and-urban-development/topics/cities-and-urban-development/city-initiatives/smart-cities_en.



A. D'Aloia, S. Scagliarini (eds.), *Law and Rights in the Smart City: The Challenge of Regulation*, The City Project 9, https://doi.org/10.1007/978-3-032-13071-6_4

tion of digital technologies within the urban context. It encompasses the efficient use of natural and economic resources, the delivery of efficient public services, the introduction of new forms of interaction between citizens and Public Administrations, the meeting of old people's needs and the protection of democratic values.

So, from the European Commission's perspective, this concept covers a bundle of initiatives devoted to efficiency and social equality that go beyond the mere technological equipment of municipalities. However, the former needs the latter to be implemented and replicated.

If, according to the European Commission, the concept of "smart city" cannot be reduced to new technologies solely, it is also true that such technologies (e.g., artificial intelligence, IoT devices, 5G) are a common trait in every urban redesign initiative. In this perspective, the technological dimension must be added to the social one in order to obtain a holistic urban renewal.

Urban initiatives are presented as the sum of six sub-components that shape civic innovation paths. These sub-components are the following:

1. smart economy;
2. smart mobility;
3. smart environment;
4. smart people;
5. smart life;
6. smart governance.

All the sub-components are strictly linked with the concept of "smartness", which is necessarily connected to the use of personal and non-personal data. Under this perspective, an intelligent city project has to be tied to the "smartness" dimension while the urban social dimension is excluded and ignored.

This second approach is incompatible with the previous one: the definitions of smart city phenomena they provide are mutually exclusive.

In order to offer an all-encompassing definition of innovative civic projects—where technology and smartness are both included, as their foundational dimensions -, a new definition of smart city should be conceptualised.

To do so, it is necessary to start again from the definitions.

What does "technological dimension" really mean? And what is the significance of "smartness"? We will try to answer both these questions in the following two paragraphs.

## 2 The Technological Essence of Smart City Projects

The foundational dimension of innovative urban projects relates to the technological endowment of civic centres firstly.

The conceptualisation of a smart city has been provided for the first time by a leading tech firm,[4] in order to demonstrate how technologies may ensure the development of urban territories in a revolutionary way. The origins of the concept suggest how tech solutions are tied to smart city initiatives: since 2008, each theorisation of innovative urban phenomena[5] has included ICT infrastructures, IoT devices, IA algorithms and data-driven solutions in their civic portrait.

Consequently, we can define technologies as one of the two essences of smart city projects. A technological dimension embraces two different aspects:

(a) the hardware/software solutions used within—or as a by-product of—the urban context;
(b) data that is collected or processed by such devices, as the so-called "urban data".

Looking at the former, hardware and software solutions (e.g., artificial intelligence, IoT devices, 5G) interact with other urban components for achieving specific objectives. These goals include increasing the efficiency and effectiveness of

[4]As we know, on sixth November, 2008 the CEO of IBM (Sam Palmisano), mentioned the *smart city* concept for the first time in his talk entitled "A Smarter Planet: The Next Leadership Agenda". On fourth November, 2011 the trademark smarter cities was officially registered as belonging to IBM.

[5]For example, Komninos N., *Intelligent cities and globalisation of innovative networks*, London 2008.

public services, the reduction of public spending, the improvement of citizens' quality of life and the development of new products and services—thanks to the adoption of data-driven decision-making processes. Additionally, we recall the Sustainable Development Goals (so-called SDGs) included in the 2030 Agenda, as defined by the UN.

In general, these objectives might be included in the so-called "twin transition": their purpose is to ensure the success of both digital and green transitions of municipal areas.

In order to achieve these objectives, hardware and software solutions might be used, for example, to increase the quality of public transport by reducing unnecessary stops[6]; to guarantee people's safety by preventing heat islands within urban areas[7]; to ensure an easier interaction between citizens and Public Administrations through the creation of civic platforms.[8]

Looking at the latter, the expression "urban data" refers to information that is collected within urban spaces through (i) public sensors,[9] (ii) residents' and tourists' personal devices,[10] or (iii) additional touchpoints.[11]

Data is often described as the "new oil": this definition refers to data-driven processes that are implemented not only in the business sector but also in the public sector. In fact, the common trait of all smart urban models is the need to collect, analyse, classify, process, store and share data collected within the metropolitan context. The importance of information for municipal initiatives suggests why we are referring to urban data as the second foundational component of the technological essence of smart city projects.

The category of urban data is also very broad.

It may include personal or non-personal data. This distinction is strictly linked with the duty to comply with Data Protection legislation (primarily, GDPR): it arises only when personal data is collected.

Urban data might also be public or private, depending on the party owning such information, i.e. Public Administration or private entities (such as enterprises, NGOs, etc.).

Such information is necessary for the development of every "intelligent" city initiative: both public or private, personal or non-personal data may address innovative urban challenges. As we know, most smart civic projects include the actions of collecting, analysing, classifying, processing, storing and sharing information (collected within the city) in order to ensure a more efficient organisation of public services and the improvement of citizens' quality of life.

However, urban data are not fixed: they change depending on the civic context in which they are generated. In other words, different cities generate different data. In our opinion, urban data is strictly tied to the characteristics of the municipality in which information is produced and collected.[12]

Unlike hardware and/or software solutions, which do not change among municipalities, urban data is shaped by the civic territory in

[6] I.e. Transport for London (TfL) uses data collected through the Oyster card system platform (which provides electronic tickets for public transports in the Greater London area) to monitor travel times, passenger numbers, incidents and other emergencies.

[7] I.e. Cagliari municipality (Italy) offers air monitoring services through an urban data platform. https://www.comune.cagliari.it/portale/page/it/cagliari_connect_the_intelligent_platform_at_city_service?contentId=VID129301

[8] I.e. Helsinki municipality has created a housing services 24/7 customer service channel to improve accessibility and customer experience and increase interaction in self-service. https://ai.hel.fi/sv/hyresbostadsbot/

[9] For example, IoT devices such as road sensors, parking sensors or smart lighting sensors.

[10] These devices are used by them for individual purposes (i.e. mobile phones, cameras, GPS devices).

[11] Like CCTV cameras, monitoring devices and other technologies that enable data collection within the city - such as urban noise mapping devices, traffic management sensors, air pollution monitoring solutions. They can be installed on guard rails, on vertical signage, on street lighting poles or in other (often busy) public places.

[12] For example, a large maritime city with a CCTV system and a small mountain town with an air monitoring system generate different types of data. Mayors and Public Administrators may launch different urban initiatives depending on information that they have.

which it originates and requires tailored projects to sustain effective technological innovations.

A one-size-fits-all solution does not work for urban data. Consequently, the technological dimension of smart city projects—which is built on this type of information—requires a unique approach for developing data strategies that perfectly fit urban characteristics.

## 3 Data and the "Smartness" Dimension of Urban Initiatives

After analysing the first foundational dimension of smart cities, we also have to investigate the concept of smartness more deeply.

"**Smartness**" is usually associated with something that is fast, intuitive or easy to use. But this correlation is reductive.

"Smartness" refers to the ability of a good or a service to simplify certain activities, to adapt itself to users' needs[13] and/or to support people in making rational decisions. Thus, key-elements of the smartness trait are (i) simplification, (ii) personalisation and (iii) support for decision-making processes.

These characteristics also recur in the smart city theorisation. The smartness of urban projects consists of the ability of such initiatives to: (i) simplify activities performed by citizens and enterprises in order to maximise their interests; (ii) personalise goods and services provided by Public Administrations through a full understanding of civic needs; (iii) support decision-making processes after investigating stakeholders' expectations.

Consequently, a municipality is smart, regardless of the size of its technological equipment, when it offers products and services that facilitate people's lives,[14] perfectly fit the needs of citizens and tourists and are designed through data-driven decision-making processes.

The overview we offered herein highlights how intimately connected technological endowment and urban smartness really are. A broad definition of intelligent civic projects must include both these foundational dimensions, and we tried to provide an all-encompassing description that embraces all of them.

The theorisation we discussed above might be applied to all urban initiatives.[15] In fact, different smart city "shapes" have arisen over time:

(a) **Smart-from-the-start cities** (i.e. India and China): these are designed from scratch and their infrastructures are built *ex novo* in order to support public strategies and for-good purposes.
(b) **Retrofitted smart cities** (i.e. U.S. and Europe): existing cities are reshaped in their structure thanks to cutting-edge technologies. In such municipalities, digital infrastructure and data analysis activities are incorporated into governance initiatives. However, these projects often apply to single urban sectors,[16] while large redesign proposals are lacking.
(c) **Social cities**: urban centres promote the use of disruptive technologies for fostering the protection of urban spaces, social activities and democratic governance through a bottom-up approach. This model is based on individual urban enhancement initiatives and may match the previous ones.

Data circulation, sharing and reuse within municipal spaces are the cornerstones of all the aforementioned urban profiles. However, cities might face different challenges depending on the original setting in which they are based.

[13] The word "smartness" can be translated as 'intelligence, effectiveness'; it shares the same etymological root as the word "sharp". A smart good or service is such if it is 'tailored' to its user and is able to respond properly to his/her needs.

[14] We recall, for example, the "15-minute city" model that implies the accessibility of essential urban services to all citizens within a 15-minute walk or bike ride. The aim of this model is to ensure that every person can benefit from "proximate" essential public services in his/her day-to-day life (see https://www.15minutecity.com/)

[15] Halegoua G. R., *Smart Cities, cit.*

[16] Such as mobility, waste management or public safety.

Ensuring access to or provision of large volumes of data can be a huge technological challenge for Public Administrators because of the absence of technological systems interoperability and the lack of practical guidelines. This challenge is particularly severe for Retrofitted smart cities in which technological equipment is adjusted over time. Conversely, Smart-from-the-start urban centres are designed for embracing broad data sharing initiatives, so they do not suffer from this challenge directly

Additionally, large volumes of data require a powerful, robust and secure urban infrastructure to be managed to avoid cyber-attacks. The probability of cyber threats increases as the number of "smart" devices grows within the city—this aspect must be considered very seriously. Cyber challenges are particularly dangerous for social cities because all the projects they promote might have a huge impact on citizens' quality of life. A cyber-attack on social cities causes the failure of important social initiatives more than for other urban models.

## 4 Data Intermediaries and Data Re-use

Decision makers must consider the challenges their cities face in order to develop a perfect urban data strategy: municipal planning initiatives depend on the vulnerabilities of each city and Public Administrations must build their interventions around them.

Additionally, the design of a good data strategy varies depending on the type of smart city we are looking at:

(a) in Smart-from-the-start cities, a data strategy is embedded in urban planning. A municipality's architecture is shaped by data-driven decisions and ICT infrastructures. In this situation, urban design is predictable and controllable;

(b) in Retrofitted smart cities, data strategies adapt themselves to urban planning. Sometimes the design of municipalities is not suitable to realise a specific data-driven civic project. Urban design follows the city's spontaneous growth more than strategic urban plans, so it appears as unpredictable and out of control;

(c) in Social cities, the data strategy is mainly focused on solving sustainability and social issues to improve citizens' quality of life. In this hypothesis, urban planning is focused on inclusion and climate-related initiatives and might include data-driven initiatives.

In conclusion, a one-size-fits-all data strategy does not exist: it changes depending on the type of smart city we have. Public Administrations must look at the urban structure before designing data-related initiatives that might improve municipalities' performance. A tailored approach ensures the satisfaction of citizens' needs and the (re)generation of data-driven public services.

It occurs through the creation of a new "urban data ecosystem" in which data strategies meet civic unique features.

After all, many urban data ecosystems have recently arisen thanks to the spread of data sharing and data reuse practices within municipal areas. An urban ecosystem has some special characteristics too.

Firstly, it is built on data protection and data governance initiatives. The former is necessary to ensure the highest protection standards to citizens' personal data and fundamental rights. The latter is necessary to choose the best-fitting sharing techniques that might be applied to data-driven smart urban projects.

Secondly, an urban data ecosystem involves all the civic stakeholders—such as citizens, Public Administration, enterprises and NGOs—and it ensures a continuous dialogue stream among them. In fact, public-private data sharing is the key element of an urban ecosystem: it allows new products and services to be created through the interaction of civic stakeholders and the exchange of data among them.

Thirdly, it requires both the establishment of specific rules for urban data sharing/reuse and the setup of legal boundaries within which these initiatives might be undertaken.

Comprehensive European legislation concerning urban data ecosystems, sharing and reuse of information in the smart city context is currently lacking. This absence poses an additional obstacle to the launch of intelligent urban projects and decreases the probability of their success.

Some sectoral provisions have been adopted since 2020. They might provide a legal framework for data-related urban initiatives until an all-encompassing mandatory discipline will be approved.

The most relevant European legal instrument is Regulation 2022/868/EU (the so-called *Data Governance Act*), which entered into force on 23rd June 2022. It has introduced the "data intermediary" figure—a subject who performs intermediation activities in exchanging information between those who may share data[17] they possess and those who are interested in using it.

This recent regulation offers a new legal framework for many data sharing and reuse initiatives within the urban context. In fact, data intermediaries perform mediating activities that allow Public Administrations to empower their data strategy while citizens' fundamental rights are protected.

For example, data intermediaries might enable bilateral or multilateral data exchanges; create data platforms that may be used by public and private stakeholders (also for not-for-profit purposes); establish secure exchange infrastructures and so on. In doing so, data intermediaries identify the right data governance and data protection setup in order to ensure safety and compliance of sharing activities, providing a degree of confidence in how such data will be used by third parties.[18]

Data intermediaries are also required to be neutral with respect to data exchange activities; to prevent frauds in the access to such data; to ensure strict information security; to allow fair, transparent and non-discriminatory access to intermediary services and to enable data exchange. In doing so, data intermediaries must look at the "best interests of data subjects".[19]

## 5 Conclusion

The legal framework described above gives smart city projects a systemic setup in which a single subject takes over the control of all urban data flows and governs them in order to improve the efficiency and efficacy of public services and thus the quality of its citizens' lives.

These provisions offer a new perspective for information sharing and reuse initiatives in the urban context: cities and enterprises might exchange huge amounts of data that allow them to provide "smart" products and services to citizens while being privacy compliant.

[17]There might be both personal and non-personal data.

[18]Whereas (22) of Regulation EU/2022/868 stresses: "Providers of data sharing services (data intermediaries) are expected to play a key role in the data economy, as a tool to facilitate the aggregation and exchange of substantial amounts of relevant data. Data intermediaries offering services that connect the different actors have the potential to contribute to the efficient pooling of data as well as to the facilitation of bilateral data sharing".

[19]Pursuant to Art. 11, DGA.

# Intelligent Systems for Intelligent Cities

Alex Buzzetti

**Abstract**

It is not easy to properly define what it means to live in an Intelligent City, given the hybrid location of this concept, straddling the material and virtual worlds. Is it possible to imagine a city in which services are increasingly interconnected with people? The role of technology, in this perspective, must change from passive to proactive. This chapter highlights the experience of Blimp, a company that operates with a set of technologies that integrate perfectly into the new lifestyles of the Intelligent City.

Living in an Intelligent City means leading a hybrid existence that weaves between the physical and virtual world, optimising the use of mental, physical and material resources. Today more than ever, to harness the future we need to anticipate it, designing and implementing highly innovative solutions in a (really) fast-moving present.

Is it possible, therefore, to imagine a city in which services are increasingly interconnected with people? We are not just talking about technologies aimed at making existing services more efficient, but, above all, about a change of logic, which transforms the role of technology: from passive to *proactive*.

If, on the one hand, technology becomes crucial to avoid unnecessary complications and efforts in daily life, on the other hand, in a highly complex context, soft skills such as agility and flexibility are key elements in the representation of contemporary citizens: indeed, they have the task of acting and moving in a sustainable way for the environment, for the city, for the society as a whole.

At **Blimp**, founded in 2017 within the **e-Novia Group**, we are operating with a set of technologies that integrate perfectly into the new lifestyles of the Intelligent City. Our company is positioned as a reference player for the collection and processing of territorial data. Our proprietary sensors, called Headcounter, measure the flows of pedestrians and vehicles thanks to computer vision and Artificial Intelligence algorithms, finding wide use in the context of the urban environment. Our systems can transfer digital metrics into the real world by collecting anonymous and aggregated data, in full compliance with privacy regulations.

An example is provided by **smart parking solutions**: through a visual signalling system, based on the actual count of parked vehicles, citizens who need to move by car can check in advance whether to park in a given area, saving one of today's most precious assets: time.

A. Buzzetti (✉)
General Manager, Blimp.ai, Milano, Italy
e-mail: alex.buzzetti@blimp.ai



A. D'Aloia, S. Scagliarini (eds.), *Law and Rights in the Smart City: The Challenge of Regulation*, The City Project 9, https://doi.org/10.1007/978-3-032-13071-6_5

We create solutions to truly improve people's quality of life when they move around the city, including by **public transport**, more sustainable from an environmental and economic point of view than a private car. We have therefore adapted our technology to optimise urban mobility, regulating waste collection based on traffic conditions; or by informing passengers waiting at bus shelters of the occupancy level of the approaching bus and informing the operations centre of the need, or not, to promptly send other vehicles.

But what are the parameters to measure how *smart* a city is? One of them is undoubtedly **security**. Our systems are suitable for monitoring the influx of people moving on foot within public spaces, preventing situations of excessive or anomalous gatherings. They can calculate the number of pedestrians, their directionality and the time spent in the areas monitored, predicting any dangerous situations and reporting them in real-time as soon as they occur.

They can also analyse road conditions in real time, monitoring vehicle flows in the main junctions even outside urban centres, based on parameters such as the count of vehicles in transit, their classification (cars, motorcycles, trucks, bus, etc.), their directionality, the speed and danger index and also any prohibited uses, such as driving against traffic or parking on the pavement.

It is therefore clear how the innovation offered by deep technologies, from sensors to Artificial Intelligence, is capable of shaping and radically transforming the world we live in, to guarantee us a future, and a present, in line with our needs.

Consider, for example, the crucial role that the **delivery** sector (which follows the same development trajectory as e-commerce) has played in the last 3 years. The recent pandemic has highlighted the multiple benefits deriving from the use of online channels—with a consequent surge in the adoption curve by companies—and changed the habits of citizens all over the world.

While on the one hand the transition to digital has greatly accelerated, on the other hand, many human interactions, including those between companies and customers, have become low-contact (and in some cases no-contact). What happens, therefore, if consumers cannot—or perhaps do not want to—test the brand experience of a company by physically visiting stores? Put simply, it is the company that moves towards them. A fast and immersive delivery is therefore the key point for building a deep relationship between people and the brand. Amazon is well aware of this, having built 100% electric drones capable of delivering orders weighing less than two and a half kilos to customers, in less than an hour from purchase. The Amazon Prime Air service has already been active since November 2022 in selected locations in California and Texas.

But it's not just the tech giants who are involved in this revolution: in a video released in 2021 to introduce and launch its international event, the Global Technology Governance Summit, the World Economic Forum included the YAPE ground drone among the three Groundbreaking Bots destined to change the world of delivery.

YAPE carries out efficient and low environmental impact last mile deliveries. In order for its use in the delivery sector to take place on a large scale, a new legislative framework needs to be defined. However, delivery robots are already today the most concrete solution to the growing demand for last mile delivery in urban areas, on the one hand because of the technological maturity now reached, on the other hand for their ability to support human operators in the most difficult and complex contexts.

Deep tech for sustainable mobility helps to design new ways to move people and goods. But what is the role of infrastructures in this context?

Today there are companies, such as **TOKBO**, which make intelligent and connected screws within a smart network. TOKBO is an Intelligent Talking Bolt Network, whose value proposition is to bring the Internet of Things to the sector of mechanical components intended for static or temporary infrastructures. The company aims to provide an effective response to the problems associated with construction scaffolding, where human error is always around the corner, and which require continuous inspections and monitoring. The same goes for bridges: 96% of Italian bridges are not adequately supervised, and surveys are limited to merely (unreliable) visual verification.

TOKBO's solution includes an Artificial Intelligence system that brings digital to the fastener industry, thanks to hardware products and maintenance services to ensure safety. Structural bolts become probes for checking clamping force, vibration and temperature, and are capable of transmitting data in real time to a digital platform that analyses all the information coming from the site. The maintenance service provides for the functional verification of the elements, also providing an indispensable predictive analysis on the tightness of the structure.

The least common multiple of these applications? It can be summed up in one word: awareness. Awareness not only of the needs of people, for whom it is crucial to move sustainably while saving time and money, but also of the role of technology, capable of proactively generating responses to the new needs of the Intelligent City: well-being, safety, quality of life.

# Participation and Smart City

Francesco Gaspari

**Abstract**

The chapter analyses what is to be understood by participation as well as the contours and framework of the smart city concept; it deals with the renewed declination of the notion of citizenship, the new participatory patterns and the repercussions on political participation and representation, highlighting how smart cities may ultimately represent—on the one hand—a new frontier for administrative law and, on the other, a privileged standpoint, not so much for the rebirth of representative democracy, which now seems to be declining, as for its overcoming.

## 1 Introductory Notes. The Meaning of "Participation"

A proper analysis of participation within a smart city entails some preliminary considerations to clarify, on the one hand, what the term "participation" means and, on the other, what we are referring to when we talk about a "smart city".

If one looks up the noun "participation" in any dictionary, one will read that the word in question originates from the Latin word *participare*, which means *make common* (*i.e.*, share in).[1] It implies a wider phrase, formed by the adjective "democratic" too,[2] and it is the opposite of the term "secrecy".

If power irresistibly tends to hide,[3] it is necessary for democracy to be designed as power in

F. Gaspari (✉)
Department of Economics and Management, Legal and Political Sciences, Guglielmo Marconi University, Rome, Italy
e-mail: f.gaspari@unimarconi.it

[1] See the Treccani Dictionary (www.treccani.it, *ad vocem*) as well as Oxford Dictionaries of English, in which the term "participation" is defined as "*Engagement with and taking part in an activity. The process through which stakeholders influence and share control over priority setting, policy-making, resource allocations, and access to public goods and services*" (www.oxfordreference.com, *ad vocem*). Another dictionary also recalls the origin of the word in question: Latin *participatio(n-)*, from *participat-* 'shared in', from the verb *participare*, based on pars, part- 'part' and capere 'take' (www.oxfordlearnersdictionaries.com, *ad vocem*). On the origin of the word in question, see also M. Nigro, *Il nodo della partecipazione*, in *Riv. trim. dir. proc. civ.*, 1980, p. 2326; F. Giglioni, S. Lariccia, *Partecipazione dei cittadini all'attività amministrativa*, in *Enc. dir.*, *Agg.*, IV, Giuffrè, Milano, 2000, pp. 943 ff., esp. p. 943; F. Cazzola, F. Lanchester, *Partecipazione e rappresentanza*, in AA. VV., *L'ambiente, l'economia e la politica*, UTET, Torino, 1987, pp. 761 ff., esp. p. 772.

[2] As has been outlined by F. Cazzola, F. Lanchester, *Partecipazione e rappresentanza*, cit., p. 772, participation is one of the aspects of the democratic method.

[3] As explained by N. Bobbio, *Democrazia e segreto*, in


A. D'Aloia, S. Scagliarini (eds.), *Law and Rights in the Smart City: The Challenge of Regulation*, The City Project 9, https://doi.org/10.1007/978-3-032-13071-6_6

public ("*potere in pubblico*"), according to the well-known definition by Norberto Bobbio,[4] who employs this expression to refer to an *active public, aware of its rights* ("*pubblico attivo, consapevole dei suoi diritti*").[5] In this perspective, "to represent" means making power visible,[6] public, exercising power in an open and transparent manner, according to the formula "*open government*",[7] so that a public opinion can form and an *effective participation* in political, economic and social life of the country can be ensured for everyone (pursuant to Article 3, par. 2, Italian Constitution).[8]

Participation, seen as *taking part* in (but also as *being part* of) may relate to both legislative activity in the strict sense and administrative activity,[9] and may take different forms, such as direct (almost like a synonym of co-decision) or indirect participation, institutionalized participation ("*partecipazione istituzionalizzata o di movimento*"), decision-oriented participation and expression-oriented participation ("*partecipazione orientata alla decisione*" and "*partecipazione orientata all'espressione*").[10]

*Participation*—which differs from *collaboration* between public administrations and private individuals[11]—is one of the key-terms by means of which the multilevel legal system has been reconfiguring relationships between public administrations and private individuals for more than 30 years, implementing the principle of democracy.[12]

In this respect, we can mention participation in administrative procedures (see Chap. III of

Id. *Teoria generale della politica,* Einaudi, Torino, 1999, pp. 352 ff., secrecy is the essence of power ("*Il segreto è l'essenza del potere*") (therein, p. 352). See also M.G. Losano, *Trasparenza e segreto: una convivenza difficile nello stato democratico*, in *Dir. pubbl.*, No. 3/2017, pp. 658 ff. now in M.G. Losano, *Scritti di informatica e diritto. Per una storia dell'informatica giuridica*, edited by P. Garbarino and M. Cavino, II, Mimesis, Milano-Udine, 2022, pp. 415 ff., spec. pp. 416 ff.

4N. Bobbio, *Democrazia e conoscenza*, in Id., *Teoria generale della politica*, cit., pp. 339 ff. With this expression, the author refers to all those institutional expedients that compel rulers to make their decisions in a transparent manner and that allow the governed to "see" how and where such decisions are taken (therein, p. 339).

5N. Bobbio, *Democrazia e conoscenza*, cit., p. 341. By the same author, see also *La democrazia e il potere invisibile*, in *Riv. it. sc. pol.*, 1980, pp. 181–203, now in *Il futuro della democrazia. In difesa delle regole del gioco*, Einaudi, Torino 1984, p. 76, as well as *Il potere occulto*, in N. Bobbio, M. Viroli, *Dialogo intorno alla repubblica*, Laterza, Bari-Roma 2001, pp. 99 ff.

6In this sense, see also C. Schmitt, *Verfassunglehre*, Duncker u. Humblot, Berlin, 1928, It. trans. by A. Caracciolo, *Dottrina della costituzione*, Giuffrè, Milano, 1984, pp. 275 ff., according to which a representation taking place in secrecy does not exist.

7See F. Costantino, *Open Government*, in *Dig. disc. pubbl., Aggiornamento*, UTET, Torino, 2015, p. 270.

8On substantial equality as a principle that also encompasses political-participatory perspectives, see A. D'Aloia, *Eguaglianza sostanziale e diritto diseguale. Contributo allo studio delle azioni positive nella prospettiva costituzionale*, Cedam, Padova, pp. 359 ff. See also G. Volpe, *Diritti politici*, in *Enc. giur.*, XI, Treccani, Roma, 1989, esp. pp. 3–4, as well as B. Pezzini, *Principio costituzionale di eguaglianza e differenza tra i sessi (A proposito della legge 125/91 sulle azioni positive)*, in *Pol. dir.*, 1993, p. 61.

9See F. Costantino, *Open Government*, cit., pp. 285–286.

10F. Cazzola, F. Lanchester, *Partecipazione e rappresentanza*, cit., pp. 772 ff.

11On the difference between participation and collaboration, see F. Costantino, *Open Government*, cit., pp. 289 ff.: the former refers to the intervention of citizens in the authoritative activity of the public administration, whereas the latter concerns supplying services (therein, p. 289). In the collaboration, which may be brought into the horizontal subsidiarity, *ex* Article 118, par. 4, It. Const., citizens are not seen as users, but as resources (therein, p. 290). On this topic, see also F. Gaspari, *The legal nature of big data and the twofold purpose of competition: in search for a new regulatory approach*, in *Dir. econ.*, No. 3/2021, pp. 239 ff.

12P. Mezzanotte, *La partecipazione alla decisione pubblica come strumento di attuazione del principio democratico. Alcune considerazioni critiche*, in *Scritti in onore di Fulco Lanchester*, II, Jovene, Napoli, 2022, pp. 893 ff.; U. Allegretti, *Democrazia partecipativa*, in *Enc. dir., Annali*, IV, Giuffrè, Milano, 2011, pp. 295 ff., esp. pp. 299 ff.; F. Giglioni, S. Lariccia, *Partecipazione dei cittadini all'attività amministrativa*, cit., pp. 943 ff. But previously see M.P. Chiti, *Partecipazione popolare e pubblica amministrazione*, Pacini, Pisa, 1977, pp. 20 ff. At European Union level, see Communication from the Commission to the European Parliament, the Council, the European Economic and Social Committee and the Committee of the Regions, COM(2021) 118, *2030 Digital Compass: the European way for the Digital Decade*, 9 March 2021, par. 3.4.

Law 7 August 1990, No. 241),[13] the different categories of access to administrative documents (within which secrecy is an exception provided by laws,[14] as occurs in the provision of Article 24 of Law No. 241/90).[15] Institutional communication,[16] and public services,[17] up until the reform of the Third Sector (with particular reference to the relationships between public administrations and active citizens, pursuant to Articles 55–57 of Legislative Decree 3 July 2017, No. 117, laying down the "Third sector Code", also referred to as TSC), together with its related topics (such as those concerning horizontal subsidiarity and the constitutional case-law on shared administration), as well as the *Government as a platform* model and new *participatory mechanisms* (above all related to local administrations), currently under investigation/being tested, including through regulatory sandboxes and living labs.

Regarding shared administration, on which we will focus more specifically in the second part of this work,[18] it should be pointed out that the

[13]On participatory institutes regulated by Law No. 241/90, the literature is very broad. Among the works of a general character, see E. Cardi, *Procedimento amministrativo*, in *Enc. giur.*, XXIV, Treccani, Roma, 1995; A. Zito, *Il procedimento amministrativo*, in F.G. Scoca (edited by), *Diritto amministrativo*, Giappichelli, Torino, 2021, pp. 207 ff.; G. Morbidelli, *Il procedimento amministrativo*, in L. Mazzarolli, G. Pericu, A. Romano, F.A. Roversi Monaco, F.G. Scoca (edited by), *Diritto amministrativo*, I, Monduzzi, Bologna, 2005, pp. 531 ff.

[14]As recalled by M. G. Losano, *Trasparenza e segreto*, cit., p. 420, in democracy secrecy is subject to the *rule of law*. But previously, in the early 1980s, Norberto Bobbio had already pointed out that in autocracy the secret of State is the rule, whereas in democracy it represents an exception provided for by laws which do not allow undue exceptions: N. Bobbio, *La democrazia e il potere invisibile*, cit., p. 94.

[15]The current legal system makes provision for three different categories of access to documents: ordinary access ("*accesso ordinario*", Articles 22 ff., Law No. 241/1990), simple civic access ("*accesso civico semplice*", set out in Article 5, par. 1, Legislative Decree 14 March 2013, No. 33) and generalised civic access ("*accesso civico generalizzato*", laid down in Article 5, par. 2, Legislative Decree No. 33/2013). On this distinction, see N. Paolantonio, *L'accesso alla documentazione amministrativa*, in F.G. Scoca (edited by), *Diritto amministrativo*, cit., pp. 265 ff. Particularly important is generalised civic access, which sets out a "*right to know*" regarding documents held by public administrations, and therefore represents an autonomous interest ("*bene della vita*") to be protected, which favours the participation of citizens, and not only an instrumental access ("*accesso strumentale*"), protected to satisfy a "final" interest ("*need to know*"). On this point, see Consiglio di Stato, Plenary Session, judgement 2 April 2020, No. 10, point 22.3, according to which in generalised civic access, the access is clearly aimed at guaranteeing democratic control over the administrative activity, in which the *right to know*, the individual interest in knowledge, is protected in itself, if and insofar as opposite reasons of public or private interest do not occur, reasons expressed by the aforementioned relative exceptions ("*eccezioni relative*") set out in Article 5 *bis*, para. 1 and 2, Legislative Decree No. 33/2013. Moreover, the Constitutional Court (judgement 21 February 2019, No. 20) underlined that the citizens' right to access information held by public administrations, based on the FOIA (Freedom of information act) model, fulfils the principle of transparency, which, as a democratic principle (Article 1, It. Const.), refers not only to all the relevant aspects of public and institutional life, but also, pursuant to Article 97, It. Const., to the efficient functioning of public administrations. See also judgements 5 April 2018, No. 69, 26 July 2018, No. 177, as well as 12 October 2017, No. 212.

[16]Institutional Communication is currently regulated by Law 7 June 2000, No. 150, as a result of a fairly long legislative process, initiated with Law 23 August 1988, No. 400, through which a strategic communication action by the Government was promoted, and the Department for information and media ("*Dipartimento per l'informazione e l'editoria*") was established, with the aim, inter alia, of fostering innovative social publicity campaigns as well as other communication activities. The purposes pursued by public administration with this form of communication are different and include, in particular, a description of the institutions' activities and their functioning, the promotion of access to public services and the knowledge of important public and social topics, the simplification of procedures, the modernization of apparatuses, the understanding of the State and of administrative procedures and, in general, the promotion of the image of Italy and its public administration in Europe and at a worldwide level. Institutional communication is an information activity through the mass media; it is directed both externally (external communication to governed individuals) and internally, within the same administration (Article 1, par. 4, Law No. 150/2000).

[17]See European Commission, COM(2021) 118, cit., par. 3.4, according to which *public services online* will "*encourage greater public participation on democratic life*".

[18]An in-depth study on this topic has been carried out by F. Gaspari, *Terzo settore ed amministrazione condivisa: profili sostanziali e processuali*", in *Società e Diritti*, No. 15/2023, pp. 71 ff.

term participation is not the same as decision/deliberation or resolution: one can participate without (necessarily) deciding.

## 2 The "Smart City" Model. Concept and Definition

The main aim of the smart city model is to enable the enhancement of the quality of life through territory, favouring this model for various purposes (first of all, housing and employment), also through the use of new technologies, so as to make urban spaces with more services (from both a quantitative and qualitative point of view) possible.

Thus, we can state that, at first sight, the smart city constitutes an expression of administrative citizenship, deemed as belonging to a territorial community[19] and no longer referring to the "ownership" of political rights, but (also) to the existence of a series of positions, which are in different ways attributable to an individual by the mere fact of being an inhabitant of a specific reality.[20]

The right to quality of life is therefore strictly linked to a relevant city or community; as a matter of fact, territory becomes—in this view of the *smart city*—instrumental in ensuring the fulfilment of the rights linked to a city or community.[21]

If the aim of the smart city model is clear, its legal framework is much less certain.[22] Indeed, it must be observed that the concept of "city" (and, *a fortiori*, that of "smart city") traditionally falls outside legal studies, which opt for other terms, such as "municipality" ("*comune*").[23] In recent years, legal research into the "city" subject has significantly increased and important studies have been carried out on this topic.[24] As has been pointed out, the "city" may be regarded as an autonomous subject of legal discipline, different from the "*comune*" (municipality), since they may be seen as two different legal realities.[25]

Similarly, also the concept of "smart city" (and that of "smart community"[26]) has been analysed by different scientific sectors and, only recently, it has drawn the attention of the legal sciences too. Indeed, this expression is mentioned even more frequently in legal acts and documents—both at domestic and at European Union levels—that identify specific policies (or "agendas") to be implemented.

In Italy, smart communities first appeared in a resolution adopted by the Cipe in 2000,[27] which

[19] R. Cavallo Perin, *La configurazione della cittadinanza amministrativa*, in *Dir. amm.*, No. 1/2004, pp. 201 ff., esp. p. 204.

[20] C.E. Gallo, *La pluralità delle cittadinanze e la cittadinanza amministrativa*, in *Dir. amm.*, No. 3/2002, pp. 481 ff., esp. p. 482.

[21] For an in-depth analysis of that view, see F. Gaspari, *Territorio, formazioni sociali e tutela dei diritti fondamentali: una nuova prospettiva del diritto alla mobilità*", Cacucci, Bari, 2021.

[22] See C. Lauri, *Smart city*, in *Dig. disc. pubbl.*, *Agg.*, VIII, UTET, Torino, 2021, pp. 378 ff.

[23] G. Piperata, *La città metropolitana: definizione, istituzione e disciplina di riferimento di un nuovo potere locale*, in A. Sterpa (edited by), *Il nuovo governo dell'area vasta. Commento alla legge 7 aprile 2014, n. 56*, Jovene, Napoli, 2014, pp. 52 ff., esp. p. 55. See also R. Cavallo Perin, *L'ordinamento giuridico della città*, in *Munus*, No. 2/2019, pp. 365 ff.

[24] Starting with an essay, now a classic, by G. Frug, *The City as a Legal Concept*, in *Harvard Law Review*, No. 93, 1979–1980, pp. 1057 ff., which was followed by numerous works, many of which are recalled in this work.

[25] R. Cavallo Perin, *Beyond the Municipality: The City, its Rights and its Rites*, in *Italian Journal of Public Law*, Issue 2, No. 2, March 2013, pp. 307 ff., esp. p. 307. See also R. Cavallo Perin, *L'ordinamento giuridico della città*, cit., pp. 369 ff. For a similar stance, see F. Giglioni, *Le città come ordinamento giuridico*, in *Ist. fed.*, No. 1/2018, pp. 29 ff., esp. pp. 70 ff., even though the author observes that compared to "*Comune*" (municipality), the term "city" does not show a clear and accurate definition, as it occurs in other circumstances in the field of law (therein, p. 74).

[26] The *smart community* concept is used when the urban reference area is wider than a city: R. Ferrara, *The smart city and the green economy in Europe: a critical approach*, in *Dir. e proc. amm.*, No. 2–3/2015, pp. 635 ff., esp. p. 636.

[27] Resolution by Cipe 21 December 2000, No. 150, laying down *Programma nazionale di ricerca 2001–2003 (decreto legislativo n. 204/1998)*. "Cipe" stands for *Comitato interministeriale per la programmazione economica* (from 1 January 2021 renamed as "Cipess", *Comitato interministeriale per la programmazione economica e lo sviluppo sostenibile*).

among other structural interventions, within the *Strategic Programme of technologies enabling the knowledge-based society* ("*Programma strategico di tecnologie abilitanti la Società della conoscenza (Ict)*"), included a programme entitled *Development of enabling platforms—smart communities* ("*Sviluppo di piattaforme abilitanti – «comunità intelligenti»*"), which, *de facto*, remained on paper.

The Italian legislator[28] and the Digital Agency for Italy ("*Agenzia per l'Italia digitale*") have referred to the expressions "smart city" and "smart community". More recently, the National Recovery and Resilience Plan ("*Piano Nazionale di Ripresa e Resilienza*"—PNRR)[29] also mentions the smart city.[30] In all such cases, no definition of smart city (or smart community) is provided.

To date, neither a legal definition of a smart city nor of a smart community exists, with the consequence that such concepts are relevant above all from a sociological point of view.

This makes "smart city" a fuzzy and multifaceted concept.[31] As a result, various attempts have been made to bring some clarity to the concept of the smart city.[32]

Some scholars outline the importance of a multidisciplinary approach to this phenomenon[33] and put forward an "integrated approach" in analysing it.[34]

Other scholars point out that in the scientific community, as well as in EU documents, *smart city* is identified as *sustainable city*,[35] while others outline that, although originally based on Information and Communication Technologies (ICTs), today smart cities go beyond that, as they represent the way sociality and lifestyle are intended, aiming at reaching post-modern versions of well-being.[36]

In this respect, some scholars observe that definitions of smart city rely on different dimensions, all to be considered, such as the digitisa-

[28] The legislator recalls—without defining it—the concept of smart communities ("*comunità intelligenti*") in Article 20, Decree Law 18 October 2012, No. 179.

[29] PNRR is a post-Pandemic recovery strategy funded through the EU Next Generation EU (NGEU) programme (see Regulation (EU) 2021/241 of the European Parliament and of the Council of 12 February 2021 establishing the Recovery and Resilience Facility). The Plan envisages a package of investments and reforms articulated into three areas and six missions. The Plan includes 53 regulatory measures, to be enacted with various sources (Decree Law, Legislative Decree, etc.): see Senate of the Republic—Studies Service Unit, Chamber of Deputies—Studies Service Unit, *Il Piano Nazionale di Ripresa e Resilienza*, Documentazione di finanza pubblica No. 28, 27 May 2021, p. 21. In literature, see E. Cavasino, *Il Piano nazionale di ripresa e resilienza e le sue fonti: dinamiche dei processi normativi in tempo di crisi*, Editoriale scientifica, Napoli, 2022. On the *governance* of the Plan see Decree Law 31 May 2021, No. 77, on *Governance del Piano nazionale di ripresa e resilienza e prime misure di rafforzamento delle strutture amministrative e di accelerazione e snellimento delle procedure*. On this Decree, see C. Di Marzio, *Il decreto semplificazioni 2021: governance del Piano nazionale di rilancio e resilienza (PNRR)*, in *Riv. amm. rep. it.*, No. 3–4/2021, pp. 117 ff.

[30] Reference to this is made under Mission 5 of the Plan (concerning *Inclusion and cohesion*), within *Investment 2.2: Integrated Urban Plans*, p. 216. On this point, see A. Aquili, *La* smart city *e le infrastrutture digitali*, in *Munus*, No. 3/2021, pp. 641 ff., esp. p. 655.

[31] See F. Gaspari, *Policy and legal challenges in implementing smart cities*, in S. Drezgić, S. Žiković, M. Tomljanović (edited by), *Smart Governments, Regions and Cities*, Proceedings of the International Scientific Conference "*Governments, Regions and Cities.* Economics of Digital Transformations", June 2–4, 2019, Rijeka, Opatija, Republic of Croatia, published by University of Rijeka, Faculty of Economics and Business, 2020, pp. 51 ff., esp. p. 52.

[32] An overview of the definitions of *smart city* can be found in Cassa depositi e prestiti, *Smart city. Progetti di sviluppo e strumenti di finanziamento*, Roma, 2013, pp. 85 ff.

[33] E. Ferrero, *Le* smart cities *nell'ordinamento giuridico*, in *Foro amm.*, 2015, pp. 1267 ff., esp. p. 1270; M. Caporale, *L'intelligenza si ripartisce o si condivide? A proposito di smartness, livelli di governo e una certa idea di città*, in *Ist. fed.*, No. 4/2015, pp. 857 ff., esp. p. 859; A. Pensi, *L'inquadramento giuridico delle «città intelligenti»*, in *www.giustamm.it*, No. 9/2015, esp. par. 1.

[34] H. Chourabi, J.R. Gil-Garcia, T.A. Pardo, T. Nam, S. Mellouli, H.J. Scholl, S. Walker, K. Nahon, *Understanding smart cities*, in *45th Hawaii International Conference on System Sciences*, 2012, available at http://observgo.uquebec.ca/observgo/fichiers/78979_B.pdf.

[35] R. Ferrara, *The smart city*, cit., p. 637. See also PNRR, cit., p. 216.

[36] G.F. Ferrari, *L'idea di smart city*, in *La prossima città*, edited by the same author, Mimesis, Milano, 2017, pp. 15–16.

tion process, the new urban governance, the development of innovative services and infrastructures, sustainable development, but also human capital, social relationships, attention to natural resources, quality of life and so-called user-centricity.[37]

Other studies and reports set parameters (or indexes) to "measure" (and classify) city smartness,[38] including the *Smart Cities Index Report*, which identifies eight parameters,[39] or *The city of the future report*, which is based on fourteen parameters.[40]

Such reports—in which competitiveness, productivity, entrepreneurial attitudes and skills take on a key role—contain sociological concepts and provide general targets and guidelines for innovative public policy development.[41]

Measuring the smartness of cities in the light of such indexes does not seem to be shareable, as it basically denies the heterogeneity of places' development, putting forward the efficiency of technical solutions applicable everywhere, in every single geographical context, even if with minor adaptations.[42]

## 3 Smart City, the Right to the City and the New Administrative Citizenship

A city is smart to the extent that it is able to increase urban well-being and quality of life,[43] so as to shape a right to the city,[44] deemed, at a legal level, as "*the individual's right to participate in a complex society, to participate in the "market" for goods and services*",[45] and as a summary of fundamental rights related to an urban territory

[37] E. Carloni, *Città intelligenti e agenda urbana: le città del futuro, il futuro delle città*, in *Munus*, No. 2/2016, pp. 240–241 (further references mentioned therein). In such terms, see also J. Eger, *Smart communities: becoming smart is not so much about developing technology as about engaging the body politic to reinvent governance in the digital age*, in *Urban Land*, 60, 2003, pp. 50 ff.; E. Ferrero, *Le* smart cities, cit., pp. 280 ff.; R. Hollands, *Will the real smart city please stand up*?, in *City*, 2008, Vol. 12, No. 3, pp. 303 ff., esp. p. 309; A. Caragliu, C. Del Bo, P. Nijkamp, *Smart cities in Europe*, in *Journal of Urban Technology*, 2011, pp. 65 ff., according to which a city is smart when "*investments in human and social capital and traditional (transport) and modern (ICT) communication infrastructure fuel sustainable economic growth and a high quality of life, with a wise management of natural resources, through participatory governance*"; M. Carley, P. Jenkins, H. Small, *Urban Development and Civil Society: The Role of Communities in Sustainable Cities*, Routledge, London, 2001, who state that social capital—deemed as "*the construction of social relations and networks of trust and reciprocity*"—is necessary so that all may participate and feel committed within smart cities.

[38] Some authors (R. Giffinger, C. Fertner, H. Kramar, R. Kalasek, N. Pichler Milanovic, E. Meijers, *Smart cities: Ranking of European medium-sized cities. Final Report*, Vienna University of Technology, 2007) identify six macro-areas of intervention within a smart city: *smart economy, smart people, smart governance, smart mobility, smart environment, smart living*.

[39] See https://smartcitiesindex.org/smartcitiesindexreport2022.

[40] The 2017 report is available at https://easyparkgroup.com/studies/cities-of-the-future/en/.

[41] See R. Ferrara, *The smart city*, cit., p. 647.

[42] See F. Gaspari, *Policy and legal challenges*, cit., p. 54.

[43] On this point, see E. Casetta, *Manuale di diritto amministrativo*, edited by F. Fracchia, Giuffrè, Milano, 2016, p. 7; F. Fracchia, P. Pantalone, *Smart City: condividere per innovare (e con il rischio di escludere?)*, in *Federalismi.it*, No. 22/2015, p. 19.

[44] J.-B. Auby, *Droit de la Ville. Du fonctionnement juridique des villes au droit à la Ville*, LexisNexis, Paris, 2013. This author sees the right to the city as a new frontier of administrative law: J.-B. Auby, *La città, nuova frontiera del diritto amministrativo?*, in *Riv. giur. urb.*, No. 1/2019, pp. 14 ff.; Id., *Per lo studio del diritto delle città*, in G. della Cananea, C. Franchini (edited by), *Il diritto che cambia*, Liber amicorum Mario Pilade Chiti, Editoriale scientifica, Napoli, 2016, pp. 205 ff. See also F. Saitta, *Il «diritto alla città»: l'attualità di una tesi antica*, in P. Stella Richter (edited by), *Ripensare la città e il suo diritto*, Giuffrè, Milano, 2022 pp. 39 ff. One of the first authors to deal with the right to the city was H. Lefebvre, *Le droit à la ville*, It. trans. by C. Bairati, *Il diritto alla città*, Marsilio, Padova, 1970, who does not, however, consider the right to the city in its technical (legal) sense, but as a right comparable to those of the well-known Declaration of Human Rights, which may be seen as the constituting act of democracy. The same author adds that such rights are never formally realised, but continuous reference is made to those rights to define the situation of the society: H. Lefebvre, *Espace et politique. Le droit à la ville II*, Anthropos, Paris, 1972, It. trans. by F. Pardi, reviewed by G. Morosato and F. Biagi, *Spazio e politica. Il diritto alla città II*, Ombrecorte, Verona, 2018, p. 131.

[45] R. Cavallo Perin, *Beyond the Municipality*, cit., p. 313; Id., *L'ordinamento giuridico della città*, cit., p. 378.

(such as, among other rights, the right to mobility, the right to security, the right to health, the right of housing),[46] and at a not strictly legal level, not only as a sustainable city (for instance, as regards energy efficiency), but also, in a wider sense, as the right to live well (happily) in the city; this also implies further relevant non-legal aspects (for instance, philosophical, political and social ones).[47]

The right to the city upheld in this work is not (only) relevant from philosophical, political and social points of view, an aspiration (or desire) to be concretely achieved through a radical criticism of the capitalist system, like the "*droit à la ville*" put forward by Lefebvre, but it takes on a specific legal dimension.[48]

Moreover, at a European Union level, there is a need to remake cities as, inter alia, centres of socio-cultural integration and foundations of democracy.[49]

The approach most capable of meeting the smart city objectives seems to be the multi-dimensional one, which starts from the premise that fundamental freedoms and rights find broad space in the urban setting,[50] and takes into account all goods and interests, beginning with the human and social dimensions, as well as the environment, digitisation, health, the dignity of life, happiness, and others.[51] However, the resulting concept of *smart city* is unitary, although it is based on different elements: a sort of "container" of all vital relationships.[52]

Even if the multilevel digital agenda is one of the components of the smart city model proposed here, technology is nevertheless called to play a pivotal—although instrumental or functional (*i.e.*, not relevant in itself)—role in terms of human progress under different aspects, such as e-health, open government, political participation, mobility and transportation (including commuting, logistics and traffic management), e-learning/e-education, the environment, security, building and construction techniques, telecommunications, waste disposal and justice (*recte*: the administration of law).[53]

In such a scenario, the smart city/community is a new way of envisioning urban life and urban well-being, and it may be deemed as the most modern manifestation of the concept of administrative citizenship ("*cittadinanza*

[46] See J.-B. Auby, *La città, nuova frontiera del diritto amministrativo?*, cit., p. 24. For a similar stance, see R. Cavallo Perin, *L'ordinamento giuridico della città*, cit., pp. 378 ff. On this point, see also F. Saitta, *Il «diritto alla città»*, cit., pp. 51–52.

[47] The approach followed here is that proposed by Henri Lefebvre in his numerous works (for an overview of it see F. Saitta, *Il «diritto alla città»*, cit., pp. 40 ff.), some of which are cited above, with the difference that in the approach followed here (discussed more widely *infra*) the right to the city is not only—in the Heideggerian sense—a call, a need, a boost to the class fight, but may be brought, under a legal perspective, into the theory of administrative citizenship, which in turn generates subjective legal situations for citizens and city users. In Heidegger, there is no eleventh *Thesis on Feuerbach*, according to which philosophers have only interpreted the world in different ways; it is, however, a matter of changing it. The *Theses on Feuerbach—Thesen über Feuerbach*—are a short work by Karl Marx, written in April 1845, and brought back to light by Friedrich Engels after the death of the author. The *Theses* were published as an annex by Engels in a work of his own in 1888.

[48] On this point, see F. Saitta, *Il «diritto alla città»*, cit., p. 63.

[49] E. Carloni, *Città intelligenti*, cit., p. 258. See also J.-B. Auby, *La città, nuova frontiera del diritto amministrativo?*, cit., p. 15; R. Cavallo Perin, *L'ordinamento giuridico della città*, cit., p. 370; G. Comazzetto, *Le città nel processo di integrazione europea*, in F. Pizzolato, G. Rivosecchi, A. Scalone (edited by), *La città oltre lo Stato*, Giappichelli, Torino, 2022, pp. 129 ff.

[50] On this approach, see J.-B. Auby, *Droit de la Ville*, cit., *passim*. Similarly, R. Cavallo Perin, *L'ordinamento giuridico della città*, cit., pp. 370, 378 ff. On such aspects, see F. Saitta, *Il «diritto alla città»*, cit., p. 51.

[51] On this point, see R. Cavallo Perin, *Beyond the Municipality*, cit., p. 310; Id., *L'ordinamento giuridico della città*, cit., pp. 378 ff.; F. Saitta, *Il «diritto alla città»*, cit., pp. 60 ff., and 65 ff.; E. Carloni, *Città intelligenti*, cit., pp. 246–247; E. Ferrero, *Le* smart cities, cit., p. 1285; F. Fracchia, P. Pantalone, *Smart City*, cit., p. 6.

[52] P. Marcuse, *Whose right(s) to what city?*, in N. Brenner, P. Marcuse, M. Mayer, *Cities for people, not for profit. Critical Urban Theory and the right to the city*, Routledge, London, New York, 2012, pp. 24 ff.

[53] A similar approach is followed by the European Commission, COM(2021) 118, cit., par. 3.4.

*amministrativa*").[54] It is a concept much wider than that of digital citizenship ("*cittadinanza digitale*") and implies the setting of a *status* concerning a specific territory, which generates subjective legal situations (both active and passive) linked to resident persons or, more accurately, to inhabitants or *denizens.*[55] In this perspective, the city goes "beyond the municipality" ("*comune*"), as it "*is made up of all those people who use it, live in it, and therefore simply constitute it as a community that is the hub of a plurality of networks*".[56] The same author clarifies that in other sciences a "community" is said to be a "flow" (a community of professionals, users, workers, transporters, etc.), and he adds that each community is an *Ordinamento giuridico* (legal system),[57] as defined by Santi Romano.[58]

Within such subjective situations, a new category of credit-related rights would be identifiable, whose content may be, for instance, a provision (of public services, also in digital mode[59]), or a participation (in local communities' decisions),[60] also in a virtual way,[61] or information (related to such rights), or forms of sharing (on which we will focus more specifically in the final part of this work), or a dialogue on (modern) uses of urban spaces or social relationships.

## 4 New Patterns of Participation in the Age of Smart Cities

Within such new patterns of administrative citizenship, and in the perspective of the smart city model followed here, participation is of paramount importance, a crucial point in what may be named as the new "right to the smart city", updating the position taken by Roberto Cavallo Perin. According to this scholar, "*the right to the city is the individual's right to participate in a complex society, to participate in the "market" for goods and services*".[62]

Participation may cover a wide range of aspects, some of which are dealt with in this work.

[54] On this view, see F. Gaspari, Smart city, *agenda urbana multilivello e nuova cittadinanza amministrativa*, Editoriale scientifica, Napoli, 2018.

[55] See R. Cavallo Perin, *La configurazione della cittadinanza amministrativa*, cit., pp. 207–208; C.E. Gallo, *La pluralità delle cittadinanze*, cit., pp. 481 ff., who notes, however, that a link with the territory obviously does not mean that the citizen as such, or rather, the person as such, is not recognized as having rights and duties over the whole of the national territory, regardless of the collectivity where that person operates or is located; this means instead that the localisation within a social formation ("*formazione sociale*") is the normal point of reference for law, given the natural and necessary sociality of human persons (therein, p. 486). Moreover, the author clarifies that, within administrative citizenship, the "*denizen*" takes the place of the "*citizen*" (therein, p. 487). Compared with the *citizen*, the *denizen* is more linked to the fact of being in a place than to the formal quality of belonging: S. Rodotà, *Tecnopolitica. La democrazia e le nuove tecnologie della comunicazione*, Laterza, Roma-Bari, 2004, p. 201. On the right to the city as a *situated right* ("*diritto situato*") in a territory, a right related to *life in a place* ("*vita nel luogo*"), see I. Agostini, *Spiragli di utopia. Lefebvre e lo spazio rurale*, in *casadellacultura.it*, 1 February 2019. For a more detailed analysis, see F. Gaspari, *Territorio,* cit., *passim.*

[56] R. Cavallo Perin, *Beyond the Municipality*, cit., p. 311; Id., *L'ordinamento giuridico della città*, cit., p. 370.

[57] R. Cavallo Perin, *Beyond the Municipality*, cit., p. 311; Id., *L'ordinamento giuridico della città*, cit., p. 372.

[58] S. Romano, *L'ordinamento giuridico*, II ed., 1946, reprinted by Giuffrè, Milano, 2013. On the city as a legal system, see also, more recently, F. Giglioni, *Le città come ordinamento giuridico*, cit., pp. 29 ff.

[59] On digital public services see G. Sgueo, *I servizi pubblici digitali*, in V. Bontempi (edited by), *Lo Stato digitale nel Piano Nazionale di Ripresa e Resilienza*, Roma Tre Press, Roma, 2022, pp. 119 ff., as well as European Commission, COM(2021) 118, cit., par. 3.4.

[60] R. Cavallo Perin, *Beyond the Municipality*, cit., p. 313.

[61] C. Chiariello, *Voto elettronico e principio di segretezza tra regola ed eccezioni*, in *Consultaonline.it*, No. 1/2023, pp. 59 ff., who refers to a right of participation in the virtual society (therein, p. 61). Under this profile, one author (L. Sartori, *Alla ricerca della* smart citizenship, in *Ist. fed.*, No. 4/2015, pp. 927 ff., esp. p. 947) has developed the concept of *smart citizenship*, pointing out that if the right to the Internet may be compared to a civil right (like that of freedom), then smart citizenship is the new social right that may attain full citizenship through new forms of participation (for instance, with *open government*).

[62] R. Cavallo Perin, *Beyond the Municipality*, cit., p. 313; Id., *L'ordinamento giuridico della città*, cit., p. 378.

Traditionally, on the constitutional side, participation is particularly important when exercising the right/duty to vote. On the administrative side, when exercising public power, participation is particularly relevant in the relationships between public administrations and private individuals (for instance, with regards to procedural participation).[63]

These are classic topics, and today they are once again gaining momentum within the smart city and, more in general, in the framework of technologies and of digitisation processes, which, thanks (*sic*!) to the pandemic, have accelerated the pace of this transformation [64] In this respect, it is interesting to note that an author, in a recent essay on new technologies and electoral procedure, refers to the image of a *building site* ("*cantiere*") that *re-opens* ("*(ri)apre*"); re-opens since the debate on the use of technology in electoral procedures dates to at least the last century.[65]

Similarly, another scholar, in a recent paper,[66] makes reference to *work in progress* ("*lavori in corso*") with regard to Artificial Intelligence, the algorithmic society and the legal dimension (*i.e.*, the role of law) in this historical phase in regulating such phenomena.

In both works, it is clear that law is—to carry the metaphor further—building its basic structure (or edifice), starting from the foundations and from the (modern) idea of the city (*i.e.*, smart city).

[63] See above, par. 1.

[64] See S. Scagliarini, *L'impiego di tecnologie digitali nel procedimento elettorale: il cantiere (ri)apre?*, in C. Panzera, A. Rauti (edited by), *Attualità di diritto pubblico*, II, Editoriale scientifica, Napoli, 2022, pp. 47 ff., esp. p. 49; C. Chiariello, *Voto elettronico*, cit., p. 75. See also Communication from the Commission to the European Parliament, the Council, the European Economic and Social Committee and the Committee of the Regions, COM(2022) 27 final, *Establishing a European Declaration on Digital rights and principles for the Digital Decade*, 26 January 2022, that outlines that the COVID-19 pandemic has accelerated the pace of digital transformation (point 2).

[65] See S. Scagliarini, *L'impiego di tecnologie digitali*, cit., p. 47.

[66] A. D'Aloia, *Intelligenza artificiale, società algoritmica, dimensione giuridica: lavori in corso*, in *Quaderni cost.*, No. 3/2022, pp. 651 ff.

## 5 Smart City and Political Participation

### 5.1 E-Democracy

If a significant number of studies about the impact of *Information and Communications Technology* (ICT) on the process of change in the public administration have been carried out and focus widely on open government, e-government, participatory and smart government, by contrast much less investigation has been undertaken into the impact of ICT on the political participation of citizens,[67] although various studies have been conducted, also dating back in time.[68]

The reference to the political participation of citizens implies, at the terminological level, a preliminary clarification of diverse concepts,[69] including those of democracy (deliberative, participatory and representative) and (political) representation.

A widespread approach in Italian literature brings "deliberative democracy"[70] into "direct democracy", considering the former as a synonym of "participatory democracy".[71] Such

[67] On this point, see A. Censorio, E-democracy: *i caratteri innovativi della democrazia in rete*, in F. Politi (edited by), *Democrazia deliberativa e rappresentanza politica: l'esperienza francese del* débat public *ed il dibattito sulla democrazia in Europa*, Giappichelli, Torino, 2022, pp. 83 ff.

[68] S. Scagliarini, *L'impiego di tecnologie digitali*, cit., p. 47.

[69] The concept of political participation is very complex, as it finds application in many fields, each of which is based on such a concept in an autonomous way. This generates doubts from a conceptual point of view, and makes it difficult to identify a precise definition. On this point, see A. Savignano, *Partecipazione politica*, in *Enc. dir.*, XXXII, Giuffrè, Milano, pp. 1 ff.

[70] On deliberative democracy, see, among numerous works, A. Floridia, *La democrazia deliberativa: teorie, processi e sistemi*, Carocci, Roma, 2013; J. Habermas, *Riflessioni sul concetto di partecipazione politica e altri saggi*, Einaudi, Torino, 1980; R. Bifulco, *Democrazia deliberativa*, in *Enc. dir.*, *Annali*, IV, Giuffrè, Milano, 2011, pp. 271 ff.

[71] See G. Colavitti, *Il "dibattito pubblico" e la partecipazione degli interessi nella prospettiva costituzionale del giusto procedimento*, in F. Politi (edited by), *Democrazia deliberativa e Rappresentanza politica*, cit., pp. 173 ff., esp. p. 174. On participatory democracy, see

expressions are frequently used in contrast with others, such as "representative democracy" or "political representation".[72]

The use of the Internet has dramatically changed the concept of political representation.[73] In recent times, new forms of democracy have arisen, such as e-democracy, to the point that some scholars wonder whether a new form of State may be identified.[74]

First of all, we have to clarify that, at a terminological level, the term "e-democracy" is not the same as "e-government".[75] *E-government*—of which *open government* is an evolution[76]—refers to all administrative and government functions by public administrations and institutions, which become more efficient and transparent, through digital technologies and the Internet.[77] In other words, *e-government* affects *how* the State comes closer to the citizens by making its services avail-

U. Allegretti, *Democrazia partecipativa*, cit., pp. 307 ff.; Id., *Basi giuridiche della democrazia partecipativa in Italia: alcuni orientamenti*, in *Dem. dir.*, No. 3/2006, pp. 151 ff.; Id., *Procedura, procedimento, processo. Un'ottica di democrazia partecipativa*, in *Dir. amm.*, No. 4/2007, pp. 779 ff.; Id., *L'amministrazione dall'attuazione costituzionale alla democrazia partecipativa*, Giuffrè, Milano, 2009; Id., *La democrazia partecipativa in Italia e in Europa*, in *Rivista AIC*, 2011, pp. 1 ff.; M. Luciani, *Democrazia rappresentativa e democrazia partecipativa*, in L. Carlassare (edited by), *La sovranità popolare nel pensiero di Esposito, Crisafulli, Livio Paladin*, Cedam, Padova, 2004, pp. 181 ff.; L. Bobbio, *Dilemmi della democrazia partecipativa*, in *Dem. dir.*, No. 4/2006, pp. 11 ff.; S. Cassese, *La partecipazione dei privati alle decisioni pubbliche. Saggio di diritto comparato*, in *Riv. trim. dir. pubbl.*, 2007, pp. 13 ff.; A. Pizzorusso, *Democrazia partecipativa e democrazia parlamentare*, in AA.VV., *Studi in onore di Antonio Amorth*, II, Giuffrè, Milano, 1982, pp. 515 ff.; W. Giulietti, *Ordinamento democratico e democrazia partecipativa: profili introduttivi*, in F. Politi (edited by), *Democrazia deliberativa e Rappresentanza politica*, cit., pp. 153 ff.

[72] See again G. Colavitti, *Il "dibattito pubblico" e la partecipazione*, cit., p. 174. See also P. Bilancia, *Democrazia diretta* versus *democrazia rappresentativa. Profili problematici nel costituzionalismo contemporaneo*, in P. Bilancia, *Crisi della rappresentanza politica nella democrazia contemporanea*, Giappichelli, Torino, 2018, pp. XI ff. On political representation, many studies exist. See, among others, F. Lanchester, *La rappresentanza in campo politico e le sue trasformazioni*, Giuffrè, Milano, 2006; F. Cazzola, F. Lanchester, *Partecipazione e rappresentanza*, cit., pp. 761 ff., to which we can also refer for the distinction between political representation and representative democracy (therein, pp. 775 ff.).

[73] A. D'Atena, *Democrazia illiberale e democrazia diretta nell'era digitale*, in *Rivista AIC*, No. 2/2019, pp. 583 ff.; A. D'Aloia, *Note sparse su costituzionalismo, sovranità, globalizzazione*, in A.P. Miras, E.C. Raffiotta, G.M. Teruel Lozano, F. Vecchio (edited by), *Sovranità e rappresentanza nell'era della globalizzazione*, Editoriale scientifica, Napoli, 2021, pp. 27 ff., esp. p. 31; G. Corso, *La rappresentanza politica nell'era digitale*, *ibidem*, pp. 175 ff.; T.E. Frosini, *Internet e democrazia*, *Ibidem*, pp. 187 ff. (therein further references on this subject by the same author); F. Costantino, *Open Government*, cit., p. 271; A. Censorio, E-democracy, cit., pp. 103–104.

[74] P. Costanzo, *La democrazia elettronica (note minime sulla cd.* e-democracy*)*, in *Dir., inf.*, No. 3/2003, pp. 465 ff., esp. p. 467, who answers negatively to the question posed, underlining that the Italian Constitution is based almost exclusively on the representative democracy formula, since direct democracy (such as, for instance, the referendum) is a symbolic and residual temperament of the representative model (therein, p. 483). In similar terms, see P. Calamandrei, *Come nasce la nuova Costituzione*, in *Il Ponte*, III, 1947 (recently re-published in *Costituzionalismo.it*, No. 3/2015, edited by G. Azzariti), who—in Constituent Assembly—upheld that parties became, *ex facto*, preparatory and complementary institutions of constitutional bodies (p. 11 of the 2015 text).

[75] P. Costanzo, *La democrazia elettronica*, cit., p. 476; L. Cuocolo, *Democrazia rappresentativa e sviluppo tecnologico*, in *Rass. parl.*, No. 4/2001, pp. 959 ff.

[76] F. Costantino, *Open Government*, cit., p. 272, according to which *open government* refers to the relationship between citizens and the public administration (therein, p. 271). In such terms, see G. Gometz, E-democracy: *forme e problemi della democrazia elettronica*, in T. Casadei, S. Pietropaoli (edited by), *Diritto e tecnologie informatiche: questioni di informatica giuridica, prospettive istituzionali e sfide sociali*, Wolters Kluwer, Milano, 2021, pp. 57 ff., esp. p. 61.

[77] A. Censorio, E-democracy, cit., p. 88. A definition of *e-government* is also provided at a European Union level. See, in particular, Communication from the Commission to the Council, the European Parliament, the European Economic and Social Committee and the Committee of the Regions, COM(2003) 567, *The Role of eGovernment for Europe's Future*, 26 September 2003, according to which *eGovernment* is defined as "*the use of information and communication technologies in public administrations combined with organisational change and new skills in order to improve public services and democratic processes and strengthen support to public policies*" (par. 3).

able.[78] *E-democracy*, instead—within whose conceptual framework all the tools of digital democracy can be found, including deliberative acts and forms of e-voting—concerns the direct power of people in the organizational sphere of the country.[79]

Therefore, *e-democracy* refers to any kind of utilisation or experimentation of digital technologies (including the Internet and mobile phones) made available by institutions to citizens so that the latter can directly participate, at administrative and political levels, in the formers' (*i.e.*, institutions') choices in every field and at every territorial level (local, national, supranational), with a view to *full and effective participation* (Article 3, par. 2, It. Const.).[80] Such participation is therefore intended not only as static and defensive, but also as dynamic (or continuous)[81] and constructive.[82]

[78]On this point see G. Gometz, E-democracy, cit., p. 61.

[79]A. Censorio, E-democracy, cit., p. 88, who points out that the *e-democracy* concept may be compared to the *e-governance* one, which makes reference to the application of information society tools to public action models, within a framework of participation and transparency in public management decisions (*ibidem*). *E-democracy* is also defined as the use of information and communication technology as a means for carrying out egalitarian procedures of self-government of people: see G. Gometz, E-democracy, cit., pp. 57 ff. On definitory aspects of *e-democracy* see also C. Chiariello, *Voto elettronico*, cit., p. 61; V. Pepe, *La democrazia elettronica tra modelli di prossimità e comunità virtuali*, in *Rass. parl.*, No 1/2021, pp. 109 ff., esp. pp. 118 ff. On the different meanings of *e-democracy* see P. Costanzo, *La democrazia elettronica*, cit., pp. 465 ff.

[80]P. Mezzanotte, *La partecipazione*, cit., p. 901, who points out the direct and general link of participation with the democratic principle and with popular sovereignty. On this point, see also A. Savignano, *Partecipazione politica*, cit., p. 11.

[81]See A. D'Aloia, *Eguaglianza sostanziale e diritto diseguale*, cit., pp. 3 ff., according to which Article 3, par. 2 of the Italian Constitution expresses a continuous (*i.e.*, endless) query (therein, p. 4). See also M. Ainis, *I soggetti deboli nella giurisprudenza costituzionale*, in *Pol. dir.*, 1999, p. 36, who points out that the implementation of constitutional rules is always partial and always has a provisional character.

[82]According to the perspective outlined by F. Benvenuti, *Il nuovo cittadino. Tra libertà garantita e libertà attiva*, Marsilio, Venezia, 1994, pp. 128–129.

*E-democracy* is a useful instrument to solve one of the thorniest questions of modern democracy, namely the way to fill the gap between one election and the following one, to end the silence of citizens, to redeem them from that condition of slavery which Rousseau referred to.[83]

However, *e-democracy*—thus understood—presents some issues,[84] the most important of which are related to its main feature, such as the fact that it (democracy) does not tolerate any representative, as upheld by Rousseau.[85] However, according to this author, direct democracy was, in practice, unrealisable, due to the difficulties of people in meeting simultaneously, as happened historically, especially during the first experiences of direct democracy in Europe.[86] Rousseau thus set the necessary conditions for democracy, namely very small States or ease of meeting for people or the possibility for all citizens to know each other.[87]

Today, the objection identified by Rousseau (and more recently echoed by other scholars) can be rebutted—with the advent of technology –, as all citizens may meet simultaneously,[88] although in a telematic way or virtually (telematic or vir-

[83]S. Rodotà, *Tecnopolitica*, cit., p. 8.

[84]On this, see A. D'Atena, *Democrazia illiberale e democrazia diretta nell'era digitale*, cit., pp. 589 ff.; P. Costanzo, *La democrazia elettronica*, cit., pp. 476–477; F. Chiusi, *Critica della democrazia digitale. La politica 2.0 alla prova dei fatti*, Codice, Torino, 2014, esp. pp. 52 ff.; A. Censorio, E-democracy, cit., pp. 102 ff.; G. Gometz, E-democracy, cit., pp. 63 ff.; M. Olivetti, *Sub Art. 1*, in R. Bifulco, A. Celotto, M. Olivetti (edited by), *Commentario alla Costituzione*, I, UTET, Torino, 2006, pp. 17 ff.; L. Elia, G. Busia, *Stato democratico*, in *Dig. disc. pubbl.*, XV, UTET, Torino, 1999, pp. 56 ff., esp. pp. 67–69.

[85]J.-J. Rousseau, *Il contratto sociale*, Laterza, Roma-Bari, 1997, p. 98.

[86]As pointed out by N. Bobbio, *Democrazia e conoscenza*, cit., p. 339, in the historic memory of European peoples, democracy shows itself for the first time through the image of the Athens *agorá*, the outdoor assembly where all citizens met to listen to the orators and then to express their opinion by raising their hand.

[87]J.-J. Rousseau, *Il contratto sociale*, cit., p. 98.

[88]On the necessity that direct democracy, duly understood, still requires today that people are physically convened to deliberate, see M. Luciani, *Democrazia rappresentativa e democrazia partecipativa*, cit., p. 182.

tual *agorá*).[89] The people are sovereign, and the same people exercise sovereignty, in the forms and within the limits set by the Constitution (Article 1, par. 2, It. Const.); hence, the constitutional order may well make provision for not (only) participatory democracy, but (also) direct democracy as a general rule.[90]

The limit of direct democracy (*e-democracy*) does not actually rest on a republican constitution's constraints,[91] but on other limits,[92] such as the culture and knowledge of those (people) called upon to express opinions. In such a scenario, according to some scholars, the direct democracy ideal is hard to achieve due to the technical and sectoral nature of rules, which the "normal" citizen is not able to understand.[93]

This limit does not appear to be insurmountable. In particular, it can be overcome, on the one hand, by increasing the quality and the quantity of information in circulation[94] and, on the other, by expanding citizens' knowledge.

Regarding the former, assuming the existence of a *right to information*, seen by the Constitutional Court as a cardinal value of the democratic system,[95] and considering that the level of democracy of a system is "measured" taking into account the portion of relevant pieces of information circulating within its boundaries, it seems evident that there is a close interrelationship between democracy-enabling conditions and technologies, which (the latter) actually enable a greater circulation of information,[96] and allow an enhancement of participation, thanks to a greater transparency of political processes, an effective exchange of information, as well as a more active participation in public life.[97]

Considering the latter (citizens' knowledge), as has been properly observed, knowledge is a necessary condition for the start of a democratic process.[98] The key issue of democracy is precisely education (*recte*: instruction),[99] and spe-

[89] Doubts are raised, among others, by M. Olivetti, *Sub Art. 1*, cit., p. 18, as well as by L. Elia, G. Busia, *Stato democratico*, cit., p. 67 (to which the phrase "*virtual agorá*" belongs). See also C. Chiariello, *Voto elettronico*, cit., p. 61, who refers to the expression "*telematic agorá*". On the fact that democracies are based on their structural physicality, and that in-person meetings are also the symbol of their existence, see F. Clementi, *Il voto "elettronico" tra deliberazione ed elezione al tempo del coronavirus*, in *Scritti in onore di Fulco Lanchester*, I, Jovene, Napoli, 2022, pp. 387 ff., esp. p. 390.

[90] On the distinction between the two forms of democracy in question, see M. Luciani, *Democrazia rappresentativa e democrazia partecipativa*, cit., pp. 181 ff.

[91] Nor does Article 49 of the It. Cost. represent a limit in this view, according to the considerations by C. Mortati, *sub Article 1*, in G. Branca (edited by), *Commentario della Costituzione. Principi fondamentali*, Vol. I, *Art. 1–12*, Zanichelli, Soc. ed. del Foro italiano, Bologna-Roma, 1975, pp. 1 ff., spec. pp. 38–39.

[92] Some scholars are of the opinion that the *interpositio* between people and political decisions is essential and cannot be deleted: see M. Luciani, *Democrazia rappresentativa e democrazia partecipativa*, cit., p. 183. Similarly, see also G. Corso, *La rappresentanza politica nell'era digitale*, cit., p. 185.

[93] L. Cuocolo, *Voto elettronico e postdemocrazia nel diritto costituzionale comparato*, in *Dir. pubbl. comp. eur.*, No. 1/2008, pp. 255 ff., esp. p. 272. In such terms, see also F. Capriglione, R. Ibrido, *La Brexit tra finanza e politica*, UTET, Torino, 2017, pp. 96 ff., esp. p. 97.

[94] In this perspective, the main issue concerns fake news, regarding which see, among others, F. Fracchia, *Le fake news come luogo di osservazione dei fenomeni della tecnologia e delle reti nella prospettiva del diritto*, in *Federalismi.it*, No. 11/2020, pp. 43 ff.; S. Sassi, *Disinformazione contro costituzionalismo*, Editoriale scientifica, Napoli, 2021; C. Valditara, *Fake news: regolamentazione e rimedi*, in *Dir. inf.*, 2021, pp. 257 ff.; C. Magnani, *Libertà di informazione online e fake news: vera emergenza? Appunti sul contrasto alla disinformazione tra legislatori statali e politiche europee*, in *Forum di Quaderni Costituzionali*, 2019; P. Passaglia, *Fake news e fake democracy: una convergenza da scongiurare*, in *Federalismi.it*, No. 11/2020, pp. 126 ff.

[95] Const. Court, judgement 10 February 1981, No. 11. On the citizen's right to complete and objective information for citizens and on the need for the continuous process of informing and training public opinion to be democratic, see Const. Court., judgement 7 May 2002, No. 155, point 2 of the grounds. On the relationship between the democratic principle and the information system, see, among others, M. Olivetti, *Sub Art. 1*, cit., p. 26.

[96] S. Rodotà, *Tecnopolitica*, cit., pp. 3 ff.

[97] C. Chiariello, *Voto elettronico*, cit., p. 62.

[98] S. Rodotà, *Tecnopolitica*, cit., p. 7.

[99] On the distinction between education and instruction, see F. Gaspari, *Istruzione, formazione e scuola digitale. Profili (critici) di diritto amministrativo*, Editoriale scientifica, Napoli, 2021, pp. 14 ff.

cifically the training of the ruling class, which should not be an inherited caste that is unreachable or accessible only through privileged channels (of various types), or a sort of oligarchy.[100]

The implementation of *e-democracy* hence implies (in general terms) a reform of the school system that is "worthy of the times", whereas the technical nature of the rules can be dealt with by preceding the deliberative moments with an informative phase,[101] taking into consideration that the availability of information has, in itself, a democratic value, as it allows transparency and diffusion of power, and may consent a control over decision-makers through an argued criticism or, even, a contrast between diverse views or hypotheses, thanks to the increasing possibility to move from pyramid decision-making models to network-decision models.[102]

This is perceived as necessary in a historical phase—like the current one—of crisis of confidence in institutions and political representation, both at domestic and European Union levels.[103]

To conclude on this point, the true dimension of *e-democracy* (including the e-vote) is that of a more direct democracy than the current one, enabling political options (*i.e.*, choices to be adopted) to be continuously subject to the opinion of people.[104] Thus, *e-democracy* enables the transition from an "*intermittent*" democracy to a "*continuous*" democracy, in which citizens can speak out at any time and in any venue, and thus become part of the daily political action.[105] Indeed, democracy does not end with the election/voting day (be it political or administrative), bringing to an end an electoral period or a decision-making procedure.[106]

Therefore, the distinction between ordinary citizens ("*cittadini ordinari*") and representative citizens ("*cittadini rappresentativi*") is also overcome, since all citizens participate directly and "among peers" ("*tra pari*"),[107] all of whom have not only the right to vote and to be elected, but also the equal right to legislate and therefore to influence the conditions determining individual status, possibly including citizen status.[108]

---

[100] P. Calamandrei, *Difendiamo la scuola democratica*, 1950, now in Id., *Costituzione e leggi di Antigone. Scritti e discorsi politici*, La Nuova Italia, Firenze, 1996, pp. 283 ff., esp. pp. 284–285.

[101] S. Rodotà, *Tecnopolitica*, cit., p. 7.

[102] S. Rodotà, *Tecnopolitica*, cit., p. 7.

[103] J.-B. Auby, *La città, nuova frontiera del diritto amministrativo?*, cit., p. 24; F. Saitta, *Il «diritto alla città»*, cit., p. 52. On the crisis and evolution of political parties, many studies have been carried out. For an introduction to this theme, see F. Lanchester, *I partiti politici tra comunità politica e istituzioni: la crisi di regime ed il riallineamento incompiuto*, in *Nomos*, No. 1/2004, pp. 99 ff.; Id. (edited by), *Dallo stato partito allo stato dei partiti. E ora?*, Cedam, Padova, 2019; Id., *Èlite dirigenziali e strumenti di selezione*, in S. Belligni (edited by), *La giraffa e il liocorno: il PCI dagli anni '70 al nuovo decennio*, Franco Angeli, Milano, 1983, pp. 151 ff., who points out that the crisis of "form of party" (or party model) is the result of the inability of the political system to recruit and select the leadership at various levels (therein, p. 151). See also, among others, P. Costanzo, *La democrazia elettronica*, cit., p. 468; F. Costantino, *Open Government*, cit., p. 271; P. Bilancia, *Crisi della democrazia rappresentativa e apertura a nuove istanze di partecipazione democratica*, in *Federalismi.it.*, No. 1/2017. Moreover, see the various chapters published in the book by P. Bilancia, *Crisi della rappresentanza politica nella democrazia contemporanea*, cit.

[104] L. Cuocolo, *Voto elettronico*, cit., p. 271.

[105] S. Rodotà, *Tecnopolitica*, cit., p. 4. By the same author, see also *Tecnologie dell'informazione e frontiere del sistema socio-politico*, in *Pol. dir.*, 1982, pp. 25 ff., esp. p. 35, in which he refers to a *continuous social contract* ("*contratto sociale continuo*"), an expression by which the author summarised the socio-political dynamics launched by the ICTs. On such aspects, see also P. Costanzo, *La democrazia elettronica*, cit., p. 475.

[106] S. Rodotà, *Tecnopolitica*, cit., p. 7. See also C. Mortati, *sub Article 1*, cit., pp. 35 ff.; A. Cariola, *Riflessioni su sovranità e rappresentanza*, in A.P. Miras, E.C. Raffiotta, G.M. Teruel Lozano, F. Vecchio (edited by), *Sovranità e rappresentanza nell'era della globalizzazione*, cit., pp. 1 ff., esp. pp. 3–4; V. Pepe, *La democrazia elettronica*, cit., p. 122.

[107] According to the view expressed by A. Pizzorno, *Introduzione allo studio della partecipazione politica*, in *Quaderni di sociologia*, No. 3–4/1966, pp. 235 ff., esp. p. 261.

[108] R. Dahrendorf, *Cittadini e partecipazione al di là della democrazia rappresentativa?*, in G. Sartori, R. Dahrendorf, *Il cittadino totale. Partecipazione eguaglianza e libertà nelle democrazie d'oggi*, in *Quaderni di biblioteca della libertà*, 3, Einaudi, Torino, 1977, pp. 33 ff., esp. pp. 45–46.

## 5.2 The Electronic Vote (e-Vote)

Within this framework, the use of technology to exercise civil rights, like the right to vote, takes on a particular meaning.

The electronic vote (e-vote) is, as mentioned above, a part of the wider category of *e-democracy*. It is not a novelty in the studies of public law in Italy, given that the debate on the e-vote in Italy dates back to 1970s.[109] The concept of the e-vote is however still far from being a precise and shared definition.[110]

Understood as distance voting, the e-vote was already under discussion at the end of the nineteenth century, when, with the extension of universal suffrage, large nations had to ensure the effective respect of the right to vote for all those that, as holders of the right in question, could not vote in their own electoral district of residence, for instance soldiers.[111]

The electronic vote may be considered as an innovative system giving the voters the possibility to express their own electoral agreement or disagreement, through the use of the most advanced technologies, also with the aim of bringing the electorate closer to a new perspective of vote expression.[112] E-voting is a key element also in the *European Commission Digital Compass* adopted in 2021,[113] especially within the process of digitisation of public services. According to the Commission, "*[s]ecured e-voting would encourage greater public participation on democratic life*".[114]

Using the *e-vote* (especially in *remote-voting* mode) provides clear advantages, not only in terms of protection of the environment, given that the current electoral procedure is based on paper documents and, therefore, implies the production of thousands of paper sheets,[115] as well as in terms of security, through the blockchain,[116] but also (and above all) because it would allow a broader democratic participation, more directed to citizens, not only for election purposes (both administrative and political), but also for important political, economic and social life decisions

[109] Important studies on this historic phase include those carried out by Vittorio Frosini and by Mario Losano. Other studies were carried out in the Eighties and afterwards. See, among others, A. Agosta, F. Lanchester, A. Spreafico (edited by), *Elezioni e automazione: tutela della regolarità del voto e prospettive di innovazione tecnologica*, Franco Angeli, Milano, 1989, and in particular, the essays by Bruno Stegagnini and by Carlo Fusaro dedicated, respectively, to electronic voting and electronic scrutiny.

[110] L. Cuocolo, *Voto elettronico*, cit., p. 257. See also C. Chiariello, *Voto elettronico*, cit., p. 63, who observes that the expression e-voting refers to voting techniques that are extremely different from each other, leading to different consequences in terms of effectivity in exercising the vote.

[111] F. Clementi, *Il voto "elettronico"*, cit., p. 392. For a broader discussion on this theme, see F. Lanchester, *Voto (diritto di), Diritto pubblico*, in *Enc. dir.*, XLVI, Giuffrè, Milano, pp. 1107 ff.

[112] L. Sergio, *Il voto elettronico nel processo di cambiamento organizzativo degli enti locali*, in *Astrid Rassegna*, No. 15/2017, p. 4. On this point, see also L. Trucco, *Il voto elettronico nella prospettiva italiana e comparata*, in *Dir. inf.*, 2011, pp. 47 ff.; G. Gometz, E-democracy, cit., p. 62; C. Chiariello, *Voto elettronico*, cit., pp. 62–63.

[113] The Digital Compass was proposed by the Commission in the form of a digital policy programme: see European Commission, COM(2021) 118, cit., par. 5.1. This programme was subsequently set up by Decision (EU) 2022/2481 of the European Parliament and of the Council of 14 December 2022 *establishing the Digital Decade Policy Programme 2030*. See Article 1 of the Decision, according to which "*[t]his Decision establishes the Digital Decade Policy Programme 2030 and sets out a monitoring and cooperation mechanism for that programme*" (par. 1).

[114] European Commission, COM(2021) 118, cit., par. 3.4.

[115] See S. Scagliarini, *L'impiego di tecnologie digitali*, cit., p. 65.

[116] See again S. Scagliarini, *L'impiego di tecnologie digitali*, cit., pp. 66 ff. See also M.F. Monterossi, *Blockchain (diritto pubblico)*, in *Dig. disc. pubbl., Aggiornamento*, VIII, UTET, Torino, 2021, pp. 29 ff., esp. p. 32.

for the country,[117] pursuant to Articles 1, 2 and 3, par. 2, It. Const.[118]

In this perspective, technology allows—alongside access to information and documents, as well as other online consultation forms[119]—citizens to be involved in the decision-making process itself, thus ensuring the realisation of the participatory-democratic ideal, by implementing the principle of maximum inclusiveness.[120]

By contrast, the e-voting system presents significant issues, especially in unattended distance (*i.e.*, remote) voting systems,[121] due to, for instance, blackout risks, hacker attacks, theft or hijacking of votes,[122] but also to "external influences" that the voter might be subjected to,[123] as well as to constraints deriving from the digital divide.[124]

This issue (digital divide) may be, at least in part, overcome by using easy and intuitive tools, as some scholars suggest.[125] However, such a thorny issue should be dealt with in a systematic and comprehensive way, as shown by the fact that it is at the heart of the European Union policies in the field of digital transition. On the one hand, the target on basic digital skills (20% of people) was established in the *Digital Education Action Plan*[126] and in the *European Pillar of Social Rights Action Plan*[127]; on the other hand, the EU Commission's Digital Compass proposed reaching a target of 20 million employed ICT specialists in the EU by 2030.[128] More recently, the digital divide has played a significant role in the rationale behind the *European Declaration on Digital Rights and Principles,*[129] an important

[117]On this point see G. Volpe, *Il costituzionalismo del Novecento*, Laterza, Roma-Bari, 2000, p. 256; C. Chiariello, *Voto elettronico*, cit., pp. 62–63.

[118]On the substantial profile of the right to vote see T. Martines, *sub Articles 56–58*, in G. Branca (edited by), *Commentario della Costituzione. Le Camere*, Vol. I, *Art. 55–63*, Zanichelli, Soc. ed. del Foro italiano, Bologna-Roma, 1984, pp. 43 ff. On the direct, widespread and differentiated participation of citizens that, based on popular sovereignty, attains its effective exercise by people, pursuant to Article 1 of the Italian Constitution, see U. Allegretti, *Democrazia partecipativa*, cit., p. 299.

[119]See L. Cuocolo, *Voto elettronico*, cit., p. 274, who refers to the case of the municipality of Madrid, which consulted the inhabitants on some regulatory options concerning planning and city management.

[120]On this, see R. Bifulco, *Democrazia deliberativa*, cit., pp. 282 ff.

[121]On the characters of both the attended and unattended vote, see A. Gratteri, *Finalità e problemi del voto elettronico: le prospettive della sua attuazione in Italia*, in *Forum di Quaderni Costituzionali—Rassegna*, 2015, par. 2, who clarifies that the attended vote is expressed at supervised, public voting places, whereas the unattended vote may be expressed by distance, in whatsoever place, using a device and without any form of supervision at the time that it is exercised. The same author, in the work in question, goes through the diverse issues linked to e-voting (therein, par. 5). On these aspects and on some e-voting issues, see also C. Chiariello, *Voto elettronico*, cit., pp. 64–65.

[122]C. Chiariello, *Voto elettronico*, cit., p. 73.

[123]Article 48, par. 2, It. Const. sets out the principle by which the vote is personal.

[124]See L. Sergio, *Il voto elettronico*, cit., pp. 11–12; P. Costanzo, *Quale partecipazione politica attraverso le nuove tecnologie comunicative in Italia*, in *Dir. inf.*, 2011, pp. 19 ff., esp. p. 29; L. Cuocolo, *Voto elettronico*, cit., p. 267; F. Capriglione, R. Ibrido, *La Brexit tra finanza e politica*, cit., pp. 96 ff.; A. Censorio, E-democracy, cit., pp. 86, 92. According to the Digital Economy and Society Index (DESI), report 2021, *Human Capital*, 84% of people in the EU in 2019 already used the Internet regularly. Nevertheless, only 56% possessed at least basic digital skills and only about one third of Europeans possessed above basic digital skills (31%).

[125]S. Scagliarini, *L'impiego di tecnologie digitali*, cit., p. 69.

[126]Communication from the Commission to the European Parliament, the Council, the European Economic and Social Committee and the Committee of the Regions, COM(2020)624 final, *The Digital Education Action Plan (2021–2027)*, 30 September 2020.

[127]Communication from the Commission to the European Parliament, the Council, the European Economic and Social Committee and the Committee of the Regions, COM(2021) 102, *The European Pillar of Social Rights Action Plan*, 3 March 2021.

[128]European Commission, COM(2021) 118, cit., par. 3.1.

[129]European Commission, COM(2022) 27, cit., point 2, in which it is pointed out that "*[w]hile the COVID-19 pandemic has accelerated the pace of this transformation, it has also increased the digital divide across the European Union*".

document despite its nature as a political declaration.[130]

Within the Digital Compass, under the "Digital Citizenship" section, the European Commission distinguishes between access to digital infrastructures and the capacity to use technologies,[131] and states that the former "*is not sufficient to define the EU's approach to its digital future*", given that "*[t]o be fully empowered, people should first have access to affordable, secure and high quality connectivity, be able to learn basic digital skills – which should become a right for all – and be equipped with other means which together allow them to fully participate in economic and societal activities of today and the future*".[132]

A further element that negatively affects the implementation of e-voting is the economic outlay to purchase equipment (like devices) and to training the staff involved in the electoral procedure (president of section, scrutineers, etc.).

Some scholars observe that this issue can also be overcome. According to this view, on the one hand, digitisation would reduce the amount of personnel involved in the procedure itself, and would allow the devices used in the procedure to be used by schools; on the other hand, the training of staff is a broader issue, that should be dealt with in a systematic and comprehensive way (similarly to that of the digital divide) by both legislators and regulators, regarding the way public administrations operate, and, therefore, it cannot be brought as a valid argument for the specific electoral procedure sector.[133]

A further issue related to the e-voting system stems from blockchain technology. Indeed, the use of blockchain in the electoral procedure, other than potential advantages in terms of security,[134] may entail significant risks to protection of the environment,[135] which takes on particular importance after the reform of articles 9 and 41 of the Italian Constitution by Constitutional Law 11 February 2022, No. 1.[136] Distributed Ledger Technologies (DLT) may indeed entail negative effects on the environment, considering that in order to make them work, a large amount of electricity is required.[137]

In cases of technology-based consent mechanisms, such as *Proof of Work*, the energy problem arises from the need to use many computers to solve the complex algorithmic operations, which, besides direct consumption, require constant, adequate ventilation to keep their temperature stable.[138]

Even a single blockchain operation needs a remarkable use of electricity. For instance, the energy consumption estimated for each transaction of Bitcoin through the blockchain amounts to about 2200.25 KWh, and namely only a little less than the equivalent amount of energy required by an Italian family in 1 year (amounting

[130] The political nature of the Declaration is clearly affirmed by the same document: "*[t]he set of digital principles that should shape Europe's digital society is proposed in the form of a joint solemn declaration of the European Parliament, the Council and the Commission*": European Commission, COM(2022) 27, cit., point 3.

[131] European Commission, COM(2021) 118, cit., par. 4. On this distinction, see F. Gaspari, Smart city, cit., p. 44.

[132] European Commission, COM(2021) 118, cit., par. 4. See also, DESI, *Human Capital*, cit., par. 1, according to which "*having an internet connection and using the internet is not sufficient; it must be paired with the appropriate skills to take advantage of the digital society*". More recently, see European Commission, COM(2022) 27, cit., point 2.

[133] S. Scagliarini, *L'impiego di tecnologie digitali*, cit., pp. 71–73.

[134] As outlined by S. Scagliarini, *L'impiego di tecnologie digitali*, cit., pp. 67–68.

[135] See F. Votta, *L'amministrazione pubblica e la tecnologia blockchain*, in *Gior. dir. amm.*, No. 4/2022, pp. 450 ff., esp. pp. 457–458.

[136] Concerning *Modifiche agli articoli 9 e 41 della Costituzione in materia di tutela dell'ambiente*.

[137] F. Votta, *L'amministrazione pubblica e la tecnologia blockchain*, cit., p. 457.

[138] K. Mohsin, *Cryptocurrency & Its Impact on Environment, Int. J. Cryp. Curr. Res.*, 1(1) (2021) pp. 1–4, available at https://papers.ssrn.com/sol3/papers.cfm?abstract_id=3846774. See also R. Filcak, R. Považan, V. Viaud, *Blockchain and the environment. An energy-intensive technology undermining climate change mitigation or a game changer for the governance of sustainability transitions?*, 2020, available at www.eea.europa.eu.

to nearly 3000 KWh).[139] As a consequence, the cryptocurrency uses more energy than entire countries such as Sweden and Malaysia [140]

This is clearly unsustainable under environmental law, also—and above all—following the energy crisis caused by the war between Russia and Ukraine, that has led the European Union[141] and Italy to adopt diverse regulatory and legislative measures,[142] and even to rename (in Italy) the Ministry of environmental transition as "Ministry of the Environment and energy security" with the establishment of the new Government in autumn 2022.

In addition to that, according to some authors, e-voting (remote unattended voting) does not appear to be compliant with the Italian Constitution.[143] In particular, to date, the unattended voting mechanism cannot guarantee the free expression of voters' intention with regard to their own political orientation. Such a mechanism does not seem to sufficiently ensure respect of the voting requirements, such as personality, freedom and secrecy.[144] The intrinsic impossibility of guaranteeing the observation of the personality and of the secrecy of voting has, to date, hindered the concrete implementation of the *home vote* (*i.e.*, remote unattended vote) in Italy, except for some local experiments.[145]

Regarding the Italian legal framework, e-voting is provided for by two regional laws: Lombardy regional law 23 February 2015, No. 3[146] and Friuli Venezia Giulia regional law 27 July 2007, No. 18.[147]

[139] F. Votta, *L'amministrazione pubblica e la tecnologia blockchain*, cit., p. 457.

[140] Cambridge Bitcoin Electricity Consumption Index (CBECI), 2021 (https://cbeci.org/). On this point, see K. Mohsin, *Cryptocurrency & Its Impact on Environment*, cit., par. 5.

[141] Economic sanctions as well as other different types of measures (such as individual sanctions, visa measures) adopted by the European Union. For an overview of measures imposed by the EU against Russia, see https://www.consilium.europa.eu/en/policies/sanctions/restrictive-measures-against-russia-over-ukraine/sanctions-against-russia-explained/. The economic sanctions, first introduced in 2014 in response to Russia's actions destabilising the situation in Ukraine, have been significantly expanded since February 2022 in light of Russia's military aggression against Ukraine: see https://www.consilium.europa.eu/en/press/press-releases/2023/01/27/russia-eu-prolongs-economic-sanctions-over-russia-s-military-aggression-against-ukraine/.

[142] At a domestic level, we refer to the following measures: Decree Law 25 February 2022, No. 14 ("*Decreto Ucraina*") converted into Law 5 April 2022, No. 28 (which also includes Decree Law 28 February 2022, No. 16, not converted into law, repealed by Article 1, par. 2, Law No. 28/2022); Decree Law 1 March 2022, No. 17 ("*Decreto Energia*") converted into Law 27 April 2022, No. 34; Decree Law 21 March 2022, No. 21 ("*Decreto Ucraina bis*") converted into Law 20 May 2022, No. 51. Further measures aimed at counteracting the negative effects of the energy crisis were subsequently adopted by the Italian Government with the following acts: Decree Law 17 May 2022, No. 50 ("*Decreto Aiuti*"), converted into Law 15 July 2022, No. 91; Decree Law 9 August 2022, No. 115 ("*Decreto Aiuti bis*"), converted into Law 21 September 2022, No. 142; Decree Law 23 September 2022, No. 144 ("*Decreto Aiuti ter*"), converted into Law 17 November 2022, No. 175; Decree Law 18 November 2022, No. 176 ("*Decreto Aiuti quarter*"), converted into Law 13 January 2023, No. 6. On these aspects, see D. Bevilacqua, *La guerra in Ucraina e le fonti energetiche: l'importanza delle politiche ambientali e le soluzioni del Green Deal europeo*, in *Riv. giur. amb. online*, No. 31/2022. With regard to the relationships between Decree Law No. 14/2022 and Decree Law No. 16/2022 and with respect to some doubts at constitutional level (both on the substantial and on procedural sides), see M. Benvenuti, *Le conseguenze costituzionali della guerra russo-ucraina. Prime considerazioni*, in *Osservatorio AIC*, No. 3/2022, p. 35.

[143] This is the conclusion reached by some authors, such as A. Gratteri, *Finalità e problemi del voto elettronico*, cit., par. 4; S. Scagliarini, *L'impiego di tecnologie digitali*, cit., pp. 80–82; C. Chiariello, *Voto elettronico*, cit., p. 65, even if with some openness towards constitutional case-law (therein, pp. 73 ff.).

[144] C. Chiariello, *Voto elettronico*, cit., p. 65.

[145] C. Chiariello, *Voto elettronico*, cit., p. 67, who recalls that the City of Avellino was the first one, at European level, to try out remote e-voting on the occasion of the constitutional referendum on 7 October 2001 (*ibidem*, note 35). On this point, see also V. Pepe, *La democrazia elettronica*, cit., p. 127.

[146] As outlined by A. Gratteri, *Finalità e problemi del voto elettronico*, cit., par. 6, the Lombardy regional law was promoted by the *Movimento 5 stelle*, and the grounds leading to the adoption of the proposal by the Regional Council mentioned cost efficiency, rapidity, greater safety and, above all, a greater involvement of voters in the decision-making process.

[147] On such regional regulations, see A. Gratteri, *Finalità e problemi del voto elettronico*, cit., para. 4 and 6.

Recently, some interesting initiatives on *e-democracy* have been identified by the National Operational Programme ("*Programma Operativo Nazionale*"—PON) Metropolitan Cities ("*Città metropolitane*") (planning 2014–2020). Within its integrated actions, the PON supports the development of instruments that are functional to e-democracy practices, aiming at improving the possibility for citizens to have access to information and to decision-making processes.

More recently, it is worth noting the form of experimentation introduced by Decree 9 July 2021, adopted by the Ministry of the Interior together with the Ministry of Technological Innovation and Digital Transition (pursuant to Law 27 December 2019, No. 160—the "Budget Law" 2020): experimentation subsequently extended to regional and administrative elections by Decree Law No. 77/2021.[148]

We are of the opinion that the time is ripe for more suitable participatory and deliberative instruments for citizens to be put in place,[149] also in the light of the provisions of the *Digital Administration Code* (*Codice dell'Amministrazione Digitale*—CAD) (in particular, Articles 3 and 9).[150]

Under such aspects, e-democracy mechanisms could be applied in electoral procedures (both attended[151] and unattended[152] votes)—thus allowing citizens to step out from what some define as a parallel reality ("*realtà parallela*") in which the world has stopped[153]—within forms of public preliminary activity ("*istruttoria pubblica*") in procedures related to the adoption of legislative or general administrative acts,[154] or the introduction of the public debate procedure[155] (typical of the French legal system[156]) for the construction of strategic public works. In addition to this, we can mention shared administration as well as the different ways of participation of both active citizens and Third sector bodies (*Enti del Terzo Settore*—ETS) (in both co-planning and co-designing phases, *ex* Article 55, TSC), on which we will focus more widely *infra*.

By the way, other countries (not only European ones)[157] resort to direct democracy instruments in

[148] On this, see S. Scagliarini, *L'impiego di tecnologie digitali*, cit., pp. 53 and 77 ff.

[149] In these terms, see also S. Scagliarini, *L'impiego di tecnologie digitali*, cit., p. 52; Id., *Tecnologie informatiche e procedimento elettorale: un matrimonio che s'ha da fare*, in *Consulta online* (Liber Amicorum per Pasquale Costanzo, I), 17 febbraio 2020, pp. 327 ff. On the opportunity of making provision for direct democracy tools by giving citizens the possibility to formulate questions, requests and petitions (within Law 7 April 2014, No. 56 and the role, closer to citizens, that the metropolitan city is called to play), see L. Fiorentino, *Il ruolo strategico dell'area vasta nella riforma dei poteri locali*, in *Riv. giur. Mezz.*, No. 4/2014, pp. 681 ff., esp. p. 687. In similar terms, see L. Cuocolo, *Voto elettronico*, cit., p. 271.

[150] See S. Scagliarini, *L'impiego di tecnologie digitali*, cit., p. 52.

[151] Both in the phase of identifying and in the phase of voting inside the ballot box. On these aspects, see S. Scagliarini, *L'impiego di tecnologie digitali*, cit., pp. 54 ff.

[152] Even if this poses doubts under a constitutional point of view, and also with regard to voter's personality/equality and secrecy/freedom, as outlined by S. Scagliarini, *L'impiego di tecnologie digitali*, cit., pp. 53 and, in particular, pp. 80–81.

[153] S. Scagliarini, *L'impiego di tecnologie digitali*, cit., p. 52.

[154] See Article 8 of the Statute of the Metropolitan City of Milan, adopted by the Metropolitan Covenant of Mayors on 22 December 2014, with Deliberation No. 2/2014 rep. gen.

[155] See Article 8, par. 6 of the Statute of the Metropolitan City of Bologna, which refers to a regulation for setting out the cases and the modes to apply the public debate ("*dibattito pubblico*"). The Statute was adopted by the Metropolitan Covenant of Mayors with Deliberation No. 1 on 23 December 2014 and was subsequently modified by Deliberation of Metropolitan Covenant No. 3 on 30 May 2016. On the "dibattito pubblico" see, among others, the book edited by F. Politi, *Democrazia deliberativa e Rappresentanza politica*, cit., *passim*; M.V. Ferroni, *Il dibattito pubblico, la democrazia partecipativa e le semplificazioni*, in *Scritti in onore di Fulco Lanchester*, I, cit., pp. 539 ff.; M.P. Vipiana, *La disciplina del dibattito pubblico nel regolamento attuativo del Codice degli appalti, tra anticipazioni regionali e suggestioni francesi*, in *Federalismi.it*, No. 2/2019; G. Pepe, *Dibattito pubblico ed infrastrutture in una prospettiva comparata*, in *Federalismi.it.*, No. 5/2019.

[156] In general, on direct participation of citizens within the urban democracy in the French system, see J.-B. Auby, *Droit de la Ville*, cit., pp. 202 ff.

[157] Attended e-voting is in operation, for instance, in the United States, Belgium, India and Brazil, whereas unattended e-voting is used, among other places, in Switzerland and Estonia. On this point, see A. Gratteri, *Finalità e problemi del voto elettronico*, cit., par. 1.

order to make important decisions, such as, for instance, in Germany, Sweden, Switzerland and Estonia.[158]

Therefore, in Italy too, the range of applications of e-democracy should definitively be broadened to allow citizens a more active participation in the political, economic and social life of the country (combined provisions of Articles 1 and 3, par. 2, Italian Constitution).

## 6 Administration and Participation in the Smart City Scenario

### 6.1 Participation and Open Government

As noted above, many studies have been investigating the *open government* topic (and previously, the *e-government* one) by analysing the impact of the *Information and Communications Technology* (ICT) on the process of change in public administrations, as regards both organisational and activity levels (for instance, digitisation, open government, participatory administration, re-use of public information, and so on, can be considered).[159]

Within the right to the smart city, that is taking shape, participation is of paramount importance with a view to re-conceptualising the city as a place of expression of an "active" political community[160] and as a fundamental element in urban transformation processes.[161]

Within *open government*, participation concerns not only the administrative procedure (provided for in Chapter III of Law No. 241/90),[162] but any activity/work by citizens provided to public powers (also within the legislation processes at different levels: not only sub-primary, but also primary and even constitutional rules).[163]

In order to ensure—at a legal level—forms of participation of citizens (as well as of city users) "*to the way of becoming power*" ("*al modo di farsi del potere*")[164] and to foster "*open democracy initiatives*",[165] the role played by administrations, especially local ones, takes on particular importance.[166]

Resorting to new technologies (but also wider reforms) has made the increasing of quality and quantity of citizens' participation possible; likewise, it has also made clear the relative risks as

[158] See F. Cazzola, F. Lanchester, *Partecipazione e rappresentanza*, cit., pp. 773–774. However, some countries backed off from using ICTs in the electoral field, due to the obstacles encountered: see S. Scagliarini, *L'impiego di tecnologie digitali*, cit., p. 49.

[159] On such aspects, many studies have been carried out. See, for all, F. Costantino, *Open Government*, cit., pp. 272 ff., as well as E. Carloni, *L'amministrazione aperta. Regole strumenti limiti dell'open government*, Maggioli, Rimini, 2014.

[160] E. Olivito, *(Dis)eguaglianza, città e periferie sociali: la prospettiva costituzionale*, in *Rivista AIC*, No. 1/2020, p. 55.

[161] F. Di Lascio, *Quali tendenze in corso nella rigenerazione delle città?*, in *Riv. giur. edil.*, 2018, p. 151.

[162] With particular reference to the administrative procedure, participating is an out-and-out general principle of the legal system: see, on such aspects, the recent analysis by Colavitti, *Il "dibattito pubblico" e la partecipazione*, cit., pp. 184 ff. See also F. Giglioni, S. Lariccia, *Partecipazione dei cittadini all'attività amministrativa*, cit., pp. 943 ff.

[163] F. Costantino, *Open Government*, cit., pp. 285–286. See also U. Allegretti, *La democrazia partecipativa in Italia e in Europa*, cit., pp. 2–3.

[164] R. Ferrara, *La partecipazione al procedimento amministrativo: un profilo critico*, in *Dir. amm.*, No. 2/2017, pp. 209 ff., esp. pp. 220 ff. (our translation); S. Cassese, *La partecipazione*, cit., pp. 14 ff.; A. Predieri, *(replica)*, in AA.VV, *Processo allo Stato. La democrazia non può permettersi il lusso di non funzionare*, Sansoni, Firenze, 1971, pp. 153 ff., esp. p. 156; F. Giglioni, S. Lariccia, *Partecipazione dei cittadini all'attività amministrativa*, cit., pp. 943 ff.; F. Gaspari, Smart city, cit., p. 169.

[165] European Commission, COM(2021) 118, cit., par. 4.

[166] On this aspect, see S. Antoniazzi, *Smart city: quadro giuridico, competenze degli enti territoriali e servizi pubblici*, in P. Stella Richter (edited by), *Ripensare la città e il suo diritto*, Giuffrè, Milano, 2022, pp. 343 ff., esp. pp. 354 ff.; J.-B. Auby, *La città, nuova frontiera del diritto amministrativo?*, cit., esp. pp. 24 and 27; F. Cortese, *Dentro il nuovo diritto delle città*, in *Munus*, No. 2/2016, p. IX; L. Giani, M. D'Orsogna, *Diritto alla città e rigenerazione urbana. Esperimenti di resilienza*, in *Scritti in onore di E. Picozza*, III, Editoriale scientifica, Napoli, 2019, pp. 2005 ff.; F. Saitta, *Il «diritto alla città»*, cit., pp. 52, 60 ff. and 65 ff.; U. Allegretti, *La democrazia partecipativa in Italia e in Europa*, cit., p. 3; V. Pepe, *La democrazia elettronica*, cit., pp. 109 ff.

well as the obstacles that hinder its full implementation (such as, for instance, the digital divide, cybersecurity, and the protection of personal data).

Different innovative ways of involvement of citizens and city users within the smart city model may be identified. Here we can, on the one hand, and in general, refer to the Third Sector code, in which forms of shared administration (co-planning and co-designing) are regulated; on the other hand, and more specifically, we can mention some regulatory options that are gaining ground in the multilevel legal system.

## 6.2 The Third Sector and the Shared Administration Model

With regard to the first point mentioned above (Third Sector), as is known, the Third Sector Code,[167] in its Title VII (Articles 55, 56 and 57), specifically makes provision for the relationships between public administrations and the Third Sector.[168]

In analysing the shared administration model, which can be merely referred to here,[169] the reflections and considerations by Feliciano Benvenuti[170]—deemed as the theorist of the principle of subsidiarity[171]—assume considerable importance. The centrality of Benvenuti's thought in theorising the shared administration model seems to be confirmed by constitutional case-law,[172] which extends the Third Sector Code to the concept of shared administration,[173] which in turn finds its own constitutional basis in the principle of horizontal subsidiarity (Article 118, par. 4, It. Const.).[174]

According to the Constitutional Court,[175] with the Code of 2017, the Italian legislator has made provision for a new collaborative relationship with public bodies; a relationship that implies the recognising that Third Sector entities (also referred to as TSEs) have a specific capacity to participate in realising the public interest together with public bodies.

[167] The Code was adopted with Legislative Decree 3 July 2017, No. 117 and entered into force on 3 August 2017 (in O.J. 18 July 2017, No. 166). The 2017 Decree sets out a new regulation of non-profit bodies and is part of a broader reform, which finds a legal basis in Law 6 June 2016, No. 106, on *Delega al Governo per la riforma del Terzo settore, dell'impresa sociale e per la disciplina del servizio civile universale*.

[168] On the Code and on the 2017 reform, see L. Gili, *Il codice del terzo settore ed i rapporti collaborativi con la p.a.*, in *Urb. app.*, No. 1/2018, pp. 15 ff.; A. Gualdani, *Il rapporto tra le pubbliche amministrazioni e gli enti del Terzo settore alla luce dei recenti interventi normativi*, in *Federalismi.it*, No. 21/2021, pp. 113 ff.; A. Mazzullo, *Il nuovo codice del terzo settore. Profili civilistici e tributari (d.lgs. 3 luglio 2017, n. 117)*, Giappichelli, Torino, 2017; L. Gori, *Terzo settore e Costituzione*, Giappichelli, Torino, 2022.

[169] For an in-depth analysis, see G. Arena, *Introduzione all'amministrazione condivisa*, in *Studi parlamentari e di politica costituzionale*, No. 117/118, 1997, pp. 29 ff.; Id., *L'amministrazione condivisa ed i suoi sviluppi nel rapporto con cittadini ed enti del Terzo Settore*, in *Giur. cost.*, 2020, p. 1453; F. Giglioni, *L'amministrazione condivisa è parte integrante della Costituzione italiana*, in *labsus.org*, 6 luglio 2020; F. Gaspari, *Terzo settore*, cit., pp. 71 ff.

[170] Among which, see at least *Il nuovo cittadino*, cit., as well as Id., *Per un diritto amministrativo paritario*, in *Studi in memoria di Enrico Guicciardi*, Cedam, Padova, 1975, pp. 807 ff., now in *Scritti giuridici*, IV, Vita e Pensiero, Milano, 2006, pp. 3223 ff.

[171] See L. Franzese, *Benvenuti, teorico del principio di sussidiarietà*, in AA.VV., *La scienza del diritto amministrativo nel pensiero di Feliciano Benvenuti*, Cedam, Padova, 2001, pp. 107 ff.

[172] Const. Court, judgement 26 June 2020, No. 131. This judgement has been widely commented by numerous scholars, among which see G. Arena, *L'amministrazione condivisa ed i suoi sviluppi*, cit., pp. 1449 ff.; E. Rossi, *Il fondamento del Terzo settore è nella Costituzione. Prime osservazioni sulla sentenza n. 131 del 2020 della Corte costituzionale*, in *Forum di Quaderni Costituzionali*, No. 3/2020, pp. 49 ff.; F. Giglioni, *L'amministrazione condivisa è parte integrante della Costituzione italiana*, cit., who points out that all the conditions are met to consider this judgement as a milestone in Italian legal history. Other works on the decision in question are cited in the present work.

[173] S.S. Scoca, *L'amministrazione condivisa nei servizi sociali: una complessa strada ancora da percorrere*, in *Dir. econ.*, No. 3/2021, pp. 83 ff., esp. p. 84.

[174] Const. Court, judgement 26 June 2020, No. 131, point 2.1 of the grounds. See also G. Arena, *L'amministrazione condivisa ed i suoi sviluppi*, cit., p. 1452.

[175] Const. Court, judgement 26 June 2020, No. 131, point 2.1 of the grounds.

In particular, pursuant to Article 55 of the TSC, a shared administration channel, alternative to that of the market, is established between public bodies and TSEs.[176] According to the Court, diverse phases of a complex procedure can be thus identified, as an expression of a different relationship between public and social private individuals/entities, not exclusively based on a synallagmatic relationship.[177]

Such phases are co-planning and co-designing and partnership (that may lead to forms of accreditation).[178]

Article 55 in question, in paragraph 1, lays down that public administrations, in carrying out their planning and organisational functions at a local level of intervention and services in the field of activity of the TSEs, must ensure the active involvement of such entities through forms of co-planning, co-designing and accreditation, put in place in compliance with the principles of Law 7 August 1990, No. 241, as well as with the provisions regulating specific procedures, and particularly those concerning local social planning ("*programmazione sociale di zona*").

The involvement of citizens undeniably leads to decision-making processes typically associated with the bottom-up approach[179]; however, this does not rule out (rather, it further reinforces) the intervention of public powers upstream (top-down approach), through specific strategic planning and through the introduction of binding rules,[180] since the mere predisposition of guidelines or recommendations or other soft law measures,[181] or acts of a merely rewarding character, are not sufficient.

## 6.3 Innovative Participatory Mechanisms

Instead, with regard to the second point indicated above, in the most recent multilevel regulatory options—in terms of involvement of citizens (and/or city users) and stakeholders in the decision-making processes of public administrations—new participatory mechanisms of various forms are emerging.

In this respect, on the one hand, we can make reference to regulatory sandboxes and living labs (for example, in the banking, financial and insurance sectors, as well as in the transportation sector); on the other hand, new participatory mechanisms have been put forward by the European Commission, such as the *Government as a platform* model within the process of digitisation of public services (also in this case, the transportation sector represents a pioneering model to take into account).

On the one hand, the recent Italian experience in regulating air mobility has shown that the role of municipalities is pivotal in terms of regional planning in urban and rural areas and the creation of dedicated infrastructure to accommodate vertiports or take-off and landing sites.[182] As recommended by the European Commission in its 2022 Drone Strategy, local administrations should be involved and be able to convey a message of certainty and transparency to society about what, how, when and where Innovative Air Mobility will be deployed.

[176] Const. Court, judgement 26 June 2020, No. 131, point 2.1 of the grounds.

[177] Const. Court, judgement 26 June 2020, No. 131, point 2.1 of the grounds.

[178] Const. Court, judgement 26 June 2020, No. 131, point 2.1 of the grounds.

[179] On cities as community creatures ("*creature delle comunità*"), see F. Giglioni, *Il diritto pubblico informale alla base della riscoperta delle città come ordinamento giuridico*, in *Riv. giur. edil.*, 2018, pp. 3 ff. On this aspect, see also F. Saitta, *Il «diritto alla città»*, cit., pp. 72–73.

[180] G. Gardini, *Alla ricerca della "città giusta". La rigenerazione come metodo di pianificazione urbana*, in *Federalismi.it.*, No. 24/2020, pp. 44 ff., esp. pp. 83–84.

[181] This is instead the opinion of C. Iaione, *La città come bene comune*, in *Aedon*, No. 1/2013 (further references are made therein), who proposes—mainly—non-binding actions and instruments (on the basis of the U.S. experience and its bottom-up approach).

[182] See the Communication from the Commission to the European Parliament, the Council, the European Economic and Social Committee and the Committee of the Regions, COM(2022) 652, *A Drone Strategy 2.0 for a smart and sustainable unmanned aircraft eco-system in Europe*, 29 November 2022, point 53.

In this regard, it is of paramount importance to allow citizens' (and city users') participation in regulatory sandboxes,[183] living labs and demonstrations when it comes to local/regional aspects to be decided regarding Innovative Air Mobility deployment.[184]

In this regard, it is worth mentioning, on the one hand, the recent Guide Lines issued by the Italian Civil Aviation Authority (ENAC) on 1 March 2023, provide for technical and procedural indications for the application of a "*new model of cooperation between the industry and the authority, in line with the orientations provided by the European Union[,] the EASA*"[185] and the International Civil Aviation Organization (ICAO),[186] as well as, on the other hand, the recent proposal to revise the Italian Airport National Plan (PNA), prepared by the same Authority (ENAC).[187] This proposal—consistently with the recent resolutions voted at the 41st ICAO Assembly[188]—explicitly provides for the involvement of local communities in order to ensure the environmental sustainability of civil aviation development, given the target of total decarbonisation by 2050.[189]

In the light of the new PNA, Urban Living labs (also within regional networks[190] or within the context of bilateral initiatives[191]) in the field of drones and Urban Air Mobility (UAM) might be designed and developed in Italy, within the wider smart city model,[192] taking into account other regional and international experiences, legal constraints, encompassing those related to civil aviation and data protection,[193] as well as, more recently, the ENAC AAM National Strategic Plan for the Development of Advanced Air Mobility in Italy ("*Piano Strategico Nazionale per lo svi-*

[183] A definition of regulatory sandboxes can be found in European Commission, *Proposal for a Regulation of the European Parliament and of the Council laying down harmonised rules on artificial intelligence (artificial intelligence act) and amending certain Union legislative acts*, COM(2021) 206 final, 21 April 2021, point 4.2.5, p. 15, according to which "*AI regulatory sandboxes establish a controlled environment to test innovative technologies for a limited time on the basis of a testing plan agreed with the competent authorities*". Regulatory sandboxes have been analysed by different authors under diverse perspectives. For an overview of these, see A. Merlino, *Regulatory Sandbox. Una nuova prospettiva ordinamentale*, Editoriale scientifica, Napoli, 2022; A. Merlino, M. Boldini, *Il diritto del fintech e la crisi delle fonti del diritto in una prospettiva comparata: il Regulatory Sandbox*, Editoriale Scientifica, Napoli, 2022; B.M. Cremona, *RegTech 3.0: verso un Regulatory Sandbox europeo?*, in *Mercato concorrenza regole*, No. 3/2019, pp. 547 ff.; E. Corapi, *Regulatory sandbox in FinTech?*, in *Dir. Commercio int.*, No. 4/2019, pp. 785 ff.

[184] See again European Commission, COM(2022) 652, cit., point 53.

[185] See ENAC Guidelines, *Sandbox ENAC*, No. 2023/001-GEN, First Edition, 1 March 2023, point 1, p. 4. A detailed definition of sandbox is provided at point 2 of the aforementioned Guidelines, which underlines its temporary character and the supervisory role of the Authority. The sandbox is compared to a *test physical or virtual laboratory* ("*laboratorio di prova fisico o virtuale*") within which it is possible to test processes, products, services, and new innovative projects.

[186] See the ICAO working paper No. HLCC 2021-WP/11. SAF/8 on *Safety Stream, Regulatory Capacity Enhancements*, presented by the Secretariat on 8 June 2021, which identifies five principles to guide the development of future regulations in the aviation sector. One of those principles is the use of Regulatory sandboxes (see point 2.2.1).

[187] The plan proposal is available at https://www.mit.gov.it/comunicazione/news/aeroporti-il-piano-nazionale-consultazione-fino-al-21-novembre-2022.

[188] Held in Montreal, Quebec, Canada, from 27 September to 7 October 2022.

[189] ENAC, Proposal of PNA, cit., p. 87.

[190] See, for instance, the *European Network of Living Labs* (https://enoll.org/network/living-labs/).

[191] See, for instance, the *urban living lab joint initiative* between India and Denmark, part of a wider Green Strategic Partnership discussed at a summit held on 28 September 2020 (https://urbanlivinglabindia.com/ and https://www.indianembassycopenhagen.gov.in/news_letter_detail/?id=3).

[192] Involving, for instance, city labs such as Torino City Lab (on which see https://www.torinocitylab.it/it/), expressly mentioned by ENAC in its National Strategic Plan (2021–2030) (point 3.3, p. 17, Figure 9) among the stakeholders to be involved. However, this list is not static, but dynamic, in the sense that it regularly evolves, and the number of stakeholders involved is expected to increase as the ecosystem matures (*ibidem*). As a consequence, other stakeholders may play a useful—if not decisive—role in such a scenario.

[193] U. Pagallo, E. Bassi, *The Governance of Unmanned Aircraft Systems (UAS): Aviation Law, Human Rights, and the Free Movement of Data in the EU*, in 30 *Minds and Machines*, 2020, p. 452.

*luppo della Mobilità Aerea Avanzata in Italia*") (2021–2030),[194] which indeed identifies some important cities among local institutions.[195]

With regard to regulatory sandboxes, the banking, financial and insurance sectors are other important areas in which this tool finds application.

The Bank of Italy defines the regulatory sandbox as "*a controlled environment where supervised entities and FinTech operators will be able to test, for a limited period of time, technologically innovative products and services in the banking, financial and insurance sectors*".[196] As clarified by the same Italian regulatory authority, the main feature of this instrument is that the tests will take place in close liaison with the supervisory authorities and can eventually benefit from a simplified transitional regime.[197]

Italian legislation (Article 36, par. 2 *bis* et seq., Decree Law 30 April 2019, No. 34) has made provision for such a regulation, delegating the implementation regulation to the Ministry of Economy and Finance. Hence, the Ministry in question has issued Decree 30 April 2021, No. 100, implementing the delegated act envisaged under the aforementioned Decree Law No. 34/2019, setting out the 'FinTech Committee rules and experimentation', *i.e* the regulatory sandbox of FinTech activities to supervisory authorities. Decree No. 100/2021 lays down a detailed regulation with regards to the FinTech Committee (Chapter I) and the FinTech experimentation (Chapter II). More recently, Article 33 of Law Decree 17 March 2023, No. 25 makes provision for measures aimed to simplify the FinTech experimentation. In particular, it lays down, inter alia, that, in compliance with the rules set out in the regulations mentioned in Article 36, para. 2 *bis* of Law Decree No. 34/2019, the Bank of Italy, the Consob and the IVASS, within their respective competences, adopt the relevant measures for the admission to the experimentation of activities under para. 2 *bis* in question together with any other related preparatory initiatives.

By participating in the sandbox, operators can test innovative products and services in constant dialogue with the supervisory authorities, also requesting that applicable regulations be waived for the purpose of the test.[198]

On the other hand, the European Commission aims at setting out new *participatory mechanisms*, and, in particular, the *Government as a platform* model in the digital public services sector.[199]

In this regard, according to the Commission, "*[b]y 2030, the EU's objective is to ensure that democratic life and public services online will be fully accessible for everyone, including persons with disabilities, and benefit from a best–in-class digital environment providing for easy-to-use, efficient and personalised services and tools with high security and privacy standards*".[200]

The tool to ensure that this objective is met is e-voting—which is expected to "*encourage greater public participation on democratic life*"—within the "*Government as a platform*", seen as "*a new way of building digital public services*". Such a platform, in the Commission's view, "*will provide a holistic and easy access to public services with a seamless interplay of*

194 The AAM Strategic Plan was published on 30 September 2021. It is available at https://www.enac.gov.it/pubblicazioni/piano-strategico-nazionale-aam-2021-2030-per-lo-sviluppo-della-mobilita-aerea-avanzata-in-italia.

195 ENAC AAM National Strategic Plan for the Development of Advanced Air Mobility in Italy, cit., Point 3.3, Figure 9.

196 See Bank of Italy, *Regulatory sandbox*, available at https://www.bancaditalia.it/focus/sandbox/index.html?com.dotmarketing.htmlpage.language=1.

197 Bank of Italy, *Regulatory sandbox*, cit., *Ibidem*.

198 Bank of Italy, *Regulatory sandbox*, cit., *Ibidem*.

199 See European Commission, COM(2021) 118, cit., par. 3.4. On the establishment of a public platform of digital services in Italy, see G. Carullo, *L'amministrazione quale piattaforma di servizi digitali*, Editoriale scientifica, Napoli, 2022. On the *Government as a platform* model, see B. Boschetti, *La transizione della pubblica amministrazione verso il modello* Government as a platform, in A. Lalli (edited by), *L'amministrazione pubblica nell'era digitale*, Giappichelli, Torino, 2022, pp. 1 ff.

200 European Commission, COM(2021) 118, cit., par. 3.4.

*advanced capabilities, such as data processing, AI and virtual reality*".[201]

The *Government as a platform* model implies some preliminary conditions to operate, including digital infrastructures. In this regard, an important role is expected to be played by cloud infrastructures as well as data and online service interoperability infrastructures. Under its Mission 1 (concerning *Digitization, Innovation, Competition, Culture and Tourism*), the Italian PNRR sets out the Cloud Public Administration (Cloud PA), in line with the Italian Cloud Strategy adopted on 7 September 2021. The Cloud PA will allow the migration of public administrations' data and IT applications to an administrations' (both central and local) cloud. This transformation will be implemented according to two complementary models, depending on data requirements with regard to performance, scalability and the sensitive nature of data. The first model gives administrations the possibility to migrate to National Strategic Cluster ("*Polo Strategico Nazionale*"—PSN), a new specific infrastructure (fully "private" or "hybrid"), located over the national territory and at the forefront of performance and security, while the second model permits them to migrate to a "public" cloud opting for one of the previously certified market operators.[202]

This new participatory model is part of "Digital Citizenship", which is being built within the EU digital society. To this end, the European Commission points out that the digital society should "*underpin and support open democracy initiatives by contributing to inclusive policy-making, enabling wide-ranging engagement with people and stimulating grass-roots action for developing local initiatives as enabler factors to improve social acceptability and public support for democratic decisions*".[203]

In this perspective, another interesting tool is the utilisation of *local participatory consultation mechanisms*, yet to be developed, but envisaged by the European Commission in its 2022 Drone Strategy.[204]

In particular, this tool, which will be funded by the Commission itself[205] and developed by EASA,[206] should ensure—in the EU Commission's view—a broad acceptance, and that comprehensive and transparent information about the technology and the type of operations involved and about the environmental impact should be provided to the people concerned. Hence, competent authorities should seek to involve them in the planned implementation process through the utilisation of local participatory consultation mechanisms.[207] The idea is that EASA develops an online platform as a "*pilot project Sustainable IAM Hub*" that would provide support to the authorities, cities, industry and other stakeholders for Innovative Air Mobility implementation.[208]

## 7 Final Remarks

From the analysis carried out so far, an advanced vision of smart city arises, within which e-democracy together with other factors related to administrative organisation and administrative action (such as, for instance, the Third Sector and the shared administration model, as well as the most recent *participatory mechanisms* of private individuals) are—once fully regulated—destined to trigger disruptive effects, at both political and legal (*i.e.*, constitutional and administrative) levels.[209]

[201] European Commission, COM(2021) 118, cit., par. 3.4.

[202] Mission 1 of the PNRR, under *Investment 1.1*, p. 92.

[203] European Commission, COM(2021) 118, cit., par. 4.

[204] European Commission, COM(2022) 652, cit., point 57.

[205] See Commission Decision of 4 July 2022 on the financing of pilot projects and preparatory actions in the field of transport for 2022, C(2022) 4509 final.

[206] EASA is the European Union Aviation Safety Agency, currently regulated by Regulation (EU) 2018/1139 of the European Parliament and of the Council of 4 July 2018 on common rules in the field of civil aviation and establishing a European Union Aviation Safety Agency (see Chapter V, Articles 75 ff.).

[207] European Commission, COM(2022) 652, cit., point 57.

[208] European Commission, COM(2022) 652, cit., point 57.

[209] M. Olivetti, *Sub Art. 1*, cit., p. 18, note 100 distinguishes the political level from the administrative one.

Here, we can focus specifically on some of them.

The first effect is the overcoming of political parties, as clearly highlighted by Feliciano Benvenuti in his numerous works.[210] The analysis of this aspect—that is still in the shadows in the studies dedicated to smart city, participation and e-democracy—can no longer be deferred following the stance taken by the Italian Constitutional Court in the well-known judgement No. 131/2020. Indeed, full adhesion to the shared administration theory leads inevitably to a severe disconnection between political parties and civil society.

This point is examined in depth by Benvenuti, according to whom true democratic participation should imply the overcoming of political parties[211] (the author also theorising an objectified administration[212]) or, in any case, returning them to their natural dimension, considering that political parties are seen today as an expression of an archaic and generally State-centred mentality, out-and-out parallel institutions.[213]

However, the overcoming of political parties would probably not solve the issues related to the party model itself and to other mediation mechanisms between people and political decisions, as the political agenda may be only partially determined from the bottom, the question submitted to people may be hetero-decided and the execution of voters' intentions, once expressed, may be manipulated or wrongly implemented or not implemented at all.[214] For instance, one may consider the possible infiltration of organised crime or other "lobbies" into the decision-making process of voter's intentions. It is known that such organisations "feed" primarily on private-public partnerships, with a decisive role that, above all, the National Anti-Bribery Authority ("*Autorità Nazionale Anticorruzione*", also referred to as ANAC) and anti-mafia documentation will be called on to play.[215] As a consequence, the capacity of the Internet to create discussion environments does not ensure, by itself, either a better decision-making process[216] or an effective and authentic democratic participation.

With the overcoming of political parties, the concept of political representation is no longer up-to-date, given that all citizens become politically active,[217] and a dilemma, that was considered to be historically archived, comes up again: that between direct democracy and representative democracy,[218] with the contextual setting aside of the parliament, considering that all decisions may be taken in real time and directly by citizens connected to the network.[219] In such a framework, it

[210] For a partially different view see M. Luciani, *Democrazia rappresentativa e democrazia partecipativa*, cit., p. 183, according to which each form of government and each technique of political decision implies the mediation of parties, groups and/or means of information.

[211] F. Benvenuti, *Il nuovo cittadino*, cit., p. 21.

[212] See F. Benvenuti, *L'amministrazione oggettivata: un nuovo modello*, in *Riv. trim. scienza amm.*, 1978, pp. 6 ff.

[213] F. Benvenuti, *Il nuovo cittadino*, cit., p. 33, as well as p. 101. On the distortive effects stemming from the pressure of parties over institutions with the necessity to reach the objectivation of power (and therefore of the State and of its administrations), see F. Benvenuti, *L'amministrazione oggettivata*, cit., pp. 6 ff.; Id., *L'ordinamento repubblicano*, Cedam, Padova, 1996; Id., *Dalla sovranità dello Stato persona alla sovranità dell'ordinamento*, in *Jus*, No. 2/1995, pp. 193 ff., now in *Scritti giuridici*, Vita e Pensiero, Milano, 2006, V, pp. 4367 ff.

[214] On such aspects see M. Luciani, *Democrazia rappresentativa e democrazia partecipativa*, cit., p. 183; A. Torre, *Chi dirige la democrazia diretta? Leviathan e Behemot, il* monstrum *bifronte della sovranità*, in P. Bilancia, *Crisi della rappresentanza politica nella democrazia contemporanea*, cit., pp. 201 ff. On hetero-directed e-democracy, see G. Gometz, E-democracy, cit., p. 65.

[215] Currently, the anti-mafia interdictory measures ("*interdittive antimafia*") are set out in Legislative Decree 6 September 2011, No. 159 (the anti-mafia Code), esp. Section II, concerning anti-mafia documentation ("*documentazione antimafia*"), as well as in Decree Law 6 November 2021, No. 152.

[216] See F. Costantino, *Open Government*, cit., p. 288.

[217] On this point, see F. Cazzola, F. Lanchester, *Partecipazione e rappresentanza*, cit., p. 776.

[218] A. D'Atena, *Democrazia illiberale e democrazia diretta nell'era digitale*, cit., p. 588. Similarly, see G. Corso, *La rappresentanza politica nell'era digitale*, cit., p. 175. See also A. Ruggeri, *"Forma di governo" e "sistema dei partiti": due categorie ormai inservibili per la teoria costituzionale?*, in *Consulta online*, No. 3/2018, pp. 606 ff.

[219] See again A. D'Atena, *Democrazia illiberale e democrazia diretta nell'era digitale*, cit., p. 588, as well as G. Corso, *La rappresentanza politica nell'era digitale*, cit., p. 175.

cannot be denied that a new form of State, no longer characterised by political parties,[220] is arising.[221]

The second effect (consequent to the first) is the resolution to what Piero Calamandrei sees as the crucial problem of democracy, namely the training of the ruling class,[222] for the simple reason that a ruling class identified as such by the political system and represented by political parties[223] is no longer identifiable. A new ruling class is indeed emerging. It is a "widespread" and unprofessional ruling class, born as a direct consequence of the absence of political parties (*recte*, a ruling class or power groups); an absence that always has a temporary character.

In this perspective, some historians point out that this absence of a ruling class can occur only in the first phase, immediately after removing the political parties (or other power groups, in a broader sense), because *élites* or oligarchies influencing the political life of a state or of a city—as happened in history—will soon arise.[224] Also in ancient history, parties (*recte*, power groups, like those established in the Greek cities and in the Roman Republic, or the medieval factions) existed, and in each political community, the fight for power led to the formation of groups in competition with each other.[225] In such a scenario, the risk is that of a hetero-directed democracy; a very high risk, considering the current Italian (and not only) cultural context.

The idea of a generational replacement of the ruling class (from a *party* one to a *widespread* one), achievable in theory, may be jeopardised by a low level of culture.

In this respect, recent studies show a growing social immobility in Italy, where poverty—not only economic, but also cultural—is inherited by children from their families of origin, thus halting intergenerational mobility brought about by a substantial inequality in access to the highest levels of schooling.[226]

This brings about severe consequences in terms of political participation[227] and in terms of generational replacement of the ruling class (not only at a political level). Indeed, as Calamandrei teaches, in a democratic system the ruling class must not be a hereditary caste, and unreachable or accessible only by way of privileged channels (of various types), a sort of oligarchy.[228]

In Calamandrei's view of democracy, in which schooling is of paramount importance, the ruling class will be open and constantly renewed by the influx of the best elements of all classes and cat-

[220]On the current form of State characterised by political parties, see F. Lanchester, *I partiti e il sistema elettorale nel pensiero di Piero Calamandrei*, in P. Barile (edited by), *Piero Calamandrei: ventidue saggi su un grande maestro*, Giuffrè, Milano, 1990, pp. 431 ff., esp. p. 452.

[221]A different conclusion is reached by P. Costanzo, *La democrazia elettronica*, cit., p. 467.

[222]The term "ruling" means those people who are able to handle and solve upcoming issues by adequately answering the questions posed by civil society: F. Lanchester, *Èlite dirigenziali e strumenti di selezione*, cit., p. 151.

[223]In Calamandrei's view, infra-party democracy was considered as the tool for a real participation of citizens within the democratic State. On this point, see F. Lanchester, *I partiti e il sistema elettorale nel pensiero di Piero Calamandrei*, cit., p. 444.

[224]A. Barbero, *La Libertà nella Storia*, lecture given at the "Festival della Comunicazione" of Camogli, September 2022. On the establishment of new democratic forms of exercising power through self-government as a typical phenomenon of the cities in the late Middle Ages see S. Cattaneo, *Città*, in *Enc. dir.*, VII, Giuffrè, Milano, 1960, par. 1.

[225]See P. Ridola, *Partiti politici*, in *Enc. dir.*, XXXII, Giuffrè, Milano,1982, pp. 66 ff., esp. p. 66. See also G. Corso, *La rappresentanza politica nell'era digitale*, cit., pp. 175 ff.

[226]M. Franzini, *Disuguaglianze inaccettabili*, Laterza, Roma-Bari, 2013. According to A. D'Aloia, *Eguaglianza sostanziale e diritto diseguale*, cit., pp. 10 ff., the substantial equality rationale is, inter alia, the emancipation and social 'mobility' of protected subjects.

[227]As occurred historically in Italy, even if there is no unanimous agreement on this, as outlined by M. Paci, *Mobilità sociale e partecipazione politica*, in *Quaderni di sociologia*, 1966, pp. 386 ff.

[228]P. Calamandrei, *Difendiamo la scuola democratica*, cit., pp. 284–285. On the problems of selection and replacement of party leadership, see also F. Lanchester, *Èlite dirigenziali e strumenti di selezione*, cit., pp. 151 ff. On the relationship between representation and responsibility and on the process leading to a shedding of responsibility of *élites* in the current constitutional system, see A. Cariola, *Riflessioni su sovranità e rappresentanza*, cit. p. 6, as well as, for a wider analysis, A. Baldassarre, *Globalizzazione contro democrazia*, Laterza, Roma-Bari, 2002, pp. 299 ff.

egories towards the top.[229] Each class and each category shall have the possibility to free its best elements upwards, because each of them may temporarily contribute to bring their work and their best personal quality to the progress of society.[230]

As Calamandrei underlines, in true democracies ("*vere democrazie*") social replacement ("*ricambio sociale*") and the mobility of the élite ("*circolazione delle élite*") (considered as the lifeblood of democracy) operate by means of renewing the ruling class.[231]

The party system has hampered all this, given that the choice does not fall to the best individuals in society, rather having to choose between political parties, programmes and "leaders" in complex fields.[232]

The overcoming of political parties, and therefore the full implementation of the democratic principle, implies a constitutional-oriented reform of the education (*recte*: instruction) system. It is indeed democracy that has to "*allow each worthy man to have his part of sun and dignity*"; and this can only be done by the schooling system, "*which is the necessary complement of the universal suffrage*".[233]

This entire political and legal programme is enshrined in Article 34 of the Italian Constitution, "*the most important article of our Constitution*"[234] and is confirmed in other articles of the Constitutional Charter (among which, 3 and 51).[235]

The third aspect (*i.e.*, disruptive effect) to take into consideration relates to the process of "building" a European Union *demos*, currently non-existent.[236]

Indeed, at political and constitutional (multi) levels, an advanced smart city model, such as the one put forward here, may fill the gap in a democracy deprived of its participatory elements and stripped of social prerequisites, with the consequent impossibility of a vital public debate, as currently occurs within the European Union.[237]

From an administrative law point of view, the smart city model outlined here fits into the perspective traced by the new administrative citizenship, from which new subjective legal situations emerge, which include not only the right of participation, but also the right of sharing.

In this respect, as Carlo Emanuele Gallo observes,[238] "*sharing values is indispensable at a*

[229] P. Calamandrei, *Difendiamo la scuola democratica*, cit., p. 285.

[230] P. Calamandrei, *Difendiamo la scuola democratica*, cit., p. 285.

[231] P. Calamandrei, *Contro il privilegio dell'istruzione*, 1946, now in Id., *Costituzione e leggi di Antigone*, cit., pp. 269 ff., esp. pp. 275 and 280.

[232] See F. Cazzola, F. Lanchester, *Partecipazione e rappresentanza*, cit., p. 777. Previously, see C. Mortati, *sub Article 1*, cit., spec. pp. 35–39.

[233] P. Calamandrei, *Difendiamo la scuola democratica*, cit., p. 286. Our translation.

[234] P. Calamandrei, *Difendiamo la scuola democratica*, cit., p. 286. Our translation.

[235] For an in-depth analysis on such aspects, see F. Gaspari, *Istruzione, formazione e scuola digitale*, cit., *passim*.

[236] D. Grimm, *The Constitution in the Process of Denationalization*, in *Constellations*, Vol. 12, No. 4/2005, pp. 447 ff.; Id., *Una Costituzione per l'Europa?*, in G. Zagrebelsky, P.P. Portinaro, J. Luther (edited by), *Il futuro della Costituzione*, Einaudi, Torino, 1996, pp. 356 ff. See also M. Luciani, *Legalità e legittimità nel processo di integrazione europea*, in G. Bonacchi (edited by), *Una costituzione senza Stato*, cit., pp. 71 ff., esp. pp. 73 ff.; G. de Vergottini, *La persistente sovranità*, cit., p. 1390.

[237] See D. Grimm, *La conquista del costituzionalismo e le sue prospettive in un mondo trasformato*, in *Scienza politica*, No. 59/2018, pp. 131 ff., esp. p. 155, according to which such aims are today hardly achievable in the European Union. On this point, see also A. D'Aloia, *Note sparse su costituzionalismo, sovranità, globalizzazione*, cit., p. 33, as well as F. Gaspari, *Il futuro dell'Unione europea tra sovranità (formale), (de)legittimazione (sostanziale) e right to exit*, in F. Gaspari (edited by), Proceedings of the Conference "*Presentazione del libro di Lorenzo Federico Pace La natura giuridica dell'Unione europea: teorie a confronto. L'Unione ai tempi della Pandemia*", held on 6 December 2021 at Università Telematica "Guglielmo Marconi" of Rome, Supplemento No. 3 of No. 3/2021 to *Rivista Trimestrale di Diritto dell'Economia*, pp. 117 ff.

[238] C.E. Gallo, *La pluralità delle cittadinanze*, cit., p. 489.

*time when, also because of the Community legal order, the concept of citizenship linked by participation in traditional forms to a collectivity of blood and territory is overcome and, instead, citizenship as an aware and voluntary acceptance of a legal framework in which one wants to live, typical of a specific collectivity increases in value*".[239]

[239]Our translation. A similar approach is followed by S. Bartole, La cittadinanza e l'identità europea, in Quad. cost., 2000, pp. 39 ff.

In conclusion, smart cities are, on the one hand, a new frontier of administrative law,[240] but also represent, on the other, a privileged standpoint, not so much for the rebirth of representative democracy,[241] which now seems to be declining, as for its overcoming.

[240]See J.-B. Auby, *La città, nuova frontiera del diritto amministrativo?*, cit., pp. 14 ff.

[241]J.-B. Auby, *La città, nuova frontiera del diritto amministrativo?*, cit., p. 24.

# From City to Smart City in Italy: A Very Long Path to Run

Edoardo Raffiotta


**Abstract**

Smart cities are an unquestionable important achievement, aspiring to protect constitutional interests, but they also imply the need for a general rethinking of the institutional and decision-making approach to the subject. The state of the art is well far away from the final goal. It is necessary to develop a comprehensive political strategy that shall, on one hand, accurately set forth solid and tangible goals and, on the other one, allocate proper resources to achieve them. It appears that the role of legal scholars might be to urge for such a plan and to stress that a better innovation might only bring about relevant benefits for the development of Italy as a country and the wellbeing of its citizens.


The landscape concerning the design, the development and, finally, the implementation of smart cities is not encouraging. Therefore, the academic debate, while always trying to adopt a practical approach, shall encourage political institutions to seize the incredible opportunity offered by new technologies.

While smart cities are already a reality in some parts of the world,[1] Italy's debate is still in its very early stages.[2] It is possible to find a few generic mentions on the matter in some institutional reports[3] or in academic research,[4] but there are just few actual digital projects which citizens can benefit from.[5]

[1]See *Smart Cities Programme*, UNDP. Available at: https://www.undp.org/policy-centre/singapore/smart-cities-programme. Siu Lon Hoe, Defining a smart nation: the case of Singapore, in Journal of Information, Communication and Ethics in Society, 2016, Vol. 14, n. 4, 323–333.

[2]G. Delle Cave, *Smart city, diritto alla città e pianificazione c.d. "intelligente"*, in federalismi.it, 2023(21), 41–80.

[3]See *Deliberazione CIPE del 21 dicembre 2000, n. 150* concerning, among others, a project for the development of smart communities.

[4]E. Ferrero, *Le smart cities nell'ordinamento giuridico*, in Foro Amministrativo, 2015(4), 1267–1286.

[5]See the project implemented by Venice, "MindCity" (https://www.mindicity.com/), an urban intelligence platform which collects data from different systems installed throughout the city and exploits them to help administrators make previsions, simulations and, ultimately, take better informed decisions.


E. Raffiotta (✉)
Università degli Studi di Milano-Bicocca,
Milan, Italy
e-mail: edoardo.raffiotta@unimib.it





A. D'Aloia, S. Scagliarini (eds.), *Law and Rights in the Smart City: The Challenge of Regulation*,
The City Project 9, https://doi.org/10.1007/978-3-032-13071-6_7

Before any further considerations, it is necessary to begin with the definition of "smart city",[6] on which there is no common ground.[7] First of all, article 114 of the Italian Constitution does not include a definition of city or municipality. The act on local municipalities (Testo Unico sugli Enti Locali, d. lgs. 267/2000) provides for a definition of municipality which does not give any contribution to the notion of smart city.[8] However, even though there is no precise definition of smart city, a quick overview of the state of the art suggests that most Italian cities lack a "smart" digital infrastructure, bringing to the light a relevant gap with other international environments.[9]

Furthermore, even when narrowing down the analysis to the sole Italian territory, it is possible to note conspicuous differences concerning the implementation of digital public services.

In the light of such differences, it appears that a better approach towards the regulation of smart cities would entail the national government as the lead policy-maker rather than having local authorities scattered throughout.

This perspective is justified for, at least, two major reasons. First of all, a minimum standard of digital services must be ensured to all citizens (without prejudice, nevertheless, to the different needs of specific municipalities). This is a scenario quite far from Italian context, where it is possible to witness a significant digital gap among local administrations: for instance, Milan is the only city that has put in place a Citizen Digital File,[10] a platform which allows citizens to manage some of their records concerning different aspects of their life, from education to tax system. These differences in the employment of basic digital services might lead to unacceptable form of discrimination among citizens.[11] Nonetheless, nowadays richer municipalities are able to implement better quality digital services than poorer and smaller town (even when it comes to basic digital services which have a significant impact on the daily life of the citizens of the local town). Thus, the allocation of the power to design and implement digital public services to local administrator does not seem to be the most reasonable and efficient solution: the State is the most suitable entity in order to ensure a minimum level of digital public services across the entire national territory.

Secondly, the smaller municipalities lack the appropriate financial and human resources to accurately carry out those digital public services that are already available.[12]

Therefore, the decision to leave the implementation of the technologies necessary for the

[6]For more in-depth attempts to define the concept of "smart city," see the definition laid out by the Italian Digital Agency (AgID - Agenzia per l'Italia al Digitale) contained in the document "Architecture for Smart Communities: conceptual vision and Recommendations to public administration" which declares: "By the term Smart City/Community (SC) we mean that place and/or territorial context where the planned and wise use of human and natural resources, appropriately managed and integrated through the many ICT technologies already available, allows the creation of an ecosystem capable of making the best use of resources and providing integrated and increasingly intelligent services (i.e. whose value is greater than the sum of the values of the parts that compose them). The axes on which the actions of an SC are developed are many: mobility, environment, energy, building quality, economy, ability to attract talent and investment, the safety of citizens and city infrastructure, and citizen participation and involvement. Prerequisites are widespread connectivity and digitization of communications and services."

[7]R. Ferrara, *The smart City and the Green Economy in Europe: a Critical Approach*, in Il Piemonte delle Autonomie, 2015(2).

[8]Article 3 par. 2 of Testo Unico sugli Enti Locali (d. lgs. 267/2000) states: "A municipality is a local authority that represents its community, looks after its interests and promotes its development".

[9]IMD World Competitiveness Center, *Smart City Index 2021*. Available at: https://imd.cld.bz/Smart-City-Index-2021.

[10]Fascicolo Digital del Cittadino.

[11]C. Napoli, *La smart city tra ambizioni europee e lacune italiane: la sfida della sostenibilità urbana* in Le Regioni, 2019(2), 445–482.

[12]G. Panattoni, *Le Smart Connected Cities: i fattori di insicurezza e (s)fiducia dei cittadini*, in Diritto pubblico 2023(2), 621–659.

realization of smart cities to the initiative of individual local authorities is not enticing. Furthermore, a system set up according to this logic would not be economically solid either, posing numerous problems of large-scale implementation of such services and, ultimately, of their inefficient distribution throughout the national territory.

Finally, even before any further legal and social consideration, there is one priority to keep in mind, which is often left out from the debate: the sophisticated digital tools, especially when used for Public Administration's activities, must meet the highest qualitative standard that ensure both the proper functioning of the software itself and the protection of all the processed information (especially, citizen's personal data). The use of software to carry out activities of public interest inevitably intertwines with fundamental principles guiding and regulating the Public Administration processes, thus posing serious risks to citizen's rights and liberties. Indeed, in a recent case, brought before the Italian Administrative Court of Appeal ("Consiglio di Stato"), the Ministry of Education had resorted to a basic software (not even to an Artificial Intelligence system) for the allocation, throughout the national territory, of teachers who had won a public selection. In the case at hand, the computer program proved to be unsuitable for goal, producing illogical and irrational outputs and forcing the Court to declare the violation of the impartiality, publicity and transparency principles.[13]

In addition, the technological tools employed by the Public Administration, considering the peculiar nature of its activities, must comply with the best available security standards, in order to protect, on the one hand, the citizens' personal data and ensure, on the other hand, the functioning of its digital infrastructure in the face of cybersecurity threats.[14]

However, the debate cannot be limited to considerations about the resilience and defense of the Public Administration's digital systems. A proactive approach is strongly necessary to develop a program that clearly defines the objectives and the proper resources to pursue them. It is worth noting, while the European Union is outlining and implementing a strategic program for the digital decade 2030 (so-called "Digital Compass")[15] that focuses on technological development, involving the Public Administration as well, a national structured plan is currently lacking.[16]

It is worth mentioning that while Europe is pushing its regulatory agenda, it is not striving to foster innovation. In recent years, Europe has implemented an ambitious program with the declared purpose of establishing its digital sovereignty: indeed, there are several sources destined to overlap in many aspects of the regulation of

[13]Cons. Stato, sez. VI, 8 aprile 2019, n. 2270.

[14]G. Alam, H. Ibrahim, *Cybersecurity implementation success factors in smart city*, in Journal of Theoretical and Applied Information Technology, 2021, Vol. 99, n. 13, 3353–3364; for a recent case of cybersecurity attack see also: D. Fadda, *Cosa ci insegna l'attacco a Westpole: cloud e ransomware priorità 2024*, in Cybersecurity360. Available at: https://www.cybersecurity360.it/cybersecurity-nazionale/cosa-ci-insegna-lattacco-a--westpole-cloud-e-ransomware-priorita-2024/. See also: G. Panattoni, *Le Smart Connected Cities: i fattori di insicurezza e (s)fiducia dei cittadini*, in Diritto pubblico 2023(2), 621–659. For national initiatives on cybersecurity see: d.l. 105/2019, *Perimetro di Sicuezza Nazionale Cibernetica*; d.l. 82/2021 establishing the national cybersecurity agency (Agenzia per la cybersicurezza nazionale).

[15]European Commission, *Communication from the Commission to the European Parliament, the Council, the European Economic and Social Committee and the Committee of the Regions - 2030 Digital Compass: the European way for the Digital Decade*, COM/2021/118 final.

[16]Indeed, there smart cities are mentioned in the Italian AI strategy, but there is no political plan that translates into specific legal provisions and that outlines the necessary financial resources to implement them.

new technological phenomenona (e.g. General Data Protection Regulation, Digital Markets Act, Digital Services Act, Artificial Intelligence Act, Data Governance Act, Data Act, etc.). However, the legislative momentum is not being followed up by corresponding efforts in terms of investment and technological development. The risk, then, could be an overregulated landscape that, indeed, lacks its very object of regulation.[17]

The data provided by the AI Index,[18] developed by Stanford University, is not encouraging and seems to confirm the marginal role of the European Union. According to the report published annually by the prestigious American university, the countries that invest the most in the development of Artificial Intelligence systems (*i.e.*, the key technology in the building real smart cities) are the United States and China, which in 2021 invested more than 70 and 20 billion, respectively, while Europe appears to be at the bottom of the ranking. A similar outcome emerges from the AI Watch Index,[19] the report drafted by the European Commission on the performance and positioning of the European Union and Member States across various dimensions of AI relevant for policymaking, that clearly indicates that there is still a lot of work to do in terms of investments, research and development.

In such a scenario, before a regulatory program, it is necessary a national political strategy. The Italian debate on the digital transformation of the Public Administration has been going on for more than twenty years (well before the enactment of the Code of Digital Administration), and it is fair to say that the principles and promises outlined at the beginning may not have had concrete implementation.

The digital transformation of the Public Administration is, first and foremost, an interest to be protected on a constitutional level.[20] The use of transparent and technologically advanced tools in Public Administration, compliant with European and national regulations (including those mentioned above), would allow the implementation of the principles of efficiency and good performance established by Article 97 of the Constitution. The use of technology, especially Artificial Intelligence, for the execution of public services could actualize the principles of impartiality and neutrality in Public Administration,[21] without necessarily relinquishing control over the specific tools. Indeed, technology must always be employed following of a human-centric approach[22]: within this context, Artificial Intelligence emerges as a useful tool to optimize resources and produce outputs that serve as guidelines, which could be either confirmed or, if necessary, discarded in a second phase by the individual.[23]

The benefits of technology, even for the Public Administration, are undeniable and the examples are quite numerous (from the increased efficiency in managing a public selection process with thousands of candidates, to the greater transparency and impartiality that could be achieved with the

[17] C. Napoli, *La smart city tra ambizioni europee e lacune italiane: la sfida della sostenibilità urbana,* ref.

[18] Daniel Zhang, Saurabh Mishra, Erik Brynjolfsson, John Etchemendy, Deep Ganguli, Barbara Grosz, Terah Lyons, James Manyika, Juan Carlos Niebles, Michael Sellitto, Yoav Shoham, Jack Clark, and Raymond Perrault, "The AI Index 2021 Annual Report," AI Index Steering Committee, Human-Centered AI Institute, Stanford University, Stanford, CA, March 2021. Available at: https://aiindex.stanford.edu/ai-index-report-2021/.

[19] Righi, R., Pineda Leon, C., Cardona, M., Soler Garrido, J., Papazoglou, M., Samoili, S. and Vazquez-Prada Baillet, M., AI Watch Index 2021, Lopez Cobo, M. and De Prato, G. editor(s), EUR 31039 EN, Publications Office of the European Union, Luxembourg, 2022, ISBN 978-92-76-53,602–4, https://doi.org/10.2760/921564, JRC128744.

[20] F. Gaspari, *Il social housing nel nuovo diritto delle città*, in federalismi.it, 2018(21), 2–32.

[21] G. Fasano, Gestione dell'algoritmo in ambito pubblico: le smart cities quale crocevia di diritti fondamentali, in federalismi.it, 2022 (29), 40–58.

[22] A. Corrado, *La trasparenza necesaria per infondere fiducia in una amministrazione algoritmica e antropocentrica*, in federalismi.it, 2023(175), 175–215.

[23] D.U. Galetta, *Digitalizzazione, Intelligenza artificiale e Pubbliche Amministrazioni: il nuovo Codice dei contratti pubblici e le sfide che ci attendono*, in federalismi.it, 2023(12), 4–14.

aid of Artificial Intelligence systems in public procurement procedures).[24]

Smart cities are an important achievement, aspiring to protect constitutional interests. To date, the state of the art is well far away from the final goal. Building upon such awareness, a comprehensive political strategy that shall, on one hand, accurately set forth solid and tangible goals and, on the other one, allocate proper resources to achieve them, is becoming always more compelling. It appears that the role of legal scholars might be to urge for such a plan and to stress that a better innovation might only bring about relevant benefits for the development of Italy as a country and the wellbeing of its citizens.

[24]Misuraca, G., and van Noordt, C.., *Overview of the use and impact of AI in public services in the EU*, EUR 30255 EN, Publications Office of the European Union, Luxembourg, 2020, ISBN 978-92-76-19540-5, https://doi.org/10.2760/039619, JRC120399.

# Automated Driving, Smart Cities and New Perspectives on Safety for Human Beings

Enrico Al Mureden

**Abstract**

Recent changes to the regulatory framework on the circulation of vehicles with assisted and highly automated driving underline the opportunity to start a reflection aimed at a profound meditation on the rules governing civil liability for road accidents. From this point of view, therefore, it appears necessary to outline a functional legal context to accelerate the transition towards new mobility models, to regulate the complex interactions between vehicles and "intelligent" structures (vehicle-to-infrastructure communication) and, at the same time, to enhance the tools which, by allowing the collection of data relating to accidents and their analysis, make it possible to achieve high levels of safety and protection of personal integrity.

## 1 From Traditional, Assisted and Automated Driving in the New EU Framework

In a context now characterised by the rapid transition from a traditional driving model focused on the fundamental role of the driver to one based on assisted driving, two significant innovations have entered into the regulatory system which, working synergistically, will contribute to making a significant change towards the progressive affirmation of Advanced Driver Assistance Systems (ADAS) and level 3 assisted driving.

14 July 2022, in fact, marks the date of entry into force of the new art. 34-bis of the Vienna Convention on road traffic of 1968, entitled "autonomous driving". The introduction of a similar provision in the UN Convention, which regulates multiple aspects of road traffic in the European Union and in a variety of adhering countries, removes the main obstacle to the

E. Al Mureden (✉)
Department of Legal Studies, University of Bologna, Bologna, Emilia-Romagna, Italy
e-mail: enrico.almureden@unibo.it



A. D'Aloia, S. Scagliarini (eds.), *Law and Rights in the Smart City: The Challenge of Regulation*, The City Project 9, https://doi.org/10.1007/978-3-032-13071-6_8

diffusion of highly automated vehicles equipped with level 3 Advanced Driver Assistance Systems,[1] i.e. cars that—while requiring the presence of a driver who carries out monitoring activities and is able to intervene if necessary —have the ability to travel independently and control every aspect connected to driving in mapped environments.[2]

The possibility of considering «satisfied the requirement of the presence of a driver in each vehicle (….) in motion» when he/she "uses an autonomous driving system compliant" with "national technical regulations, and with any international legal instrument" (art, 34-bis, letter a, Vienna Convention), or "to the national legislation that regulates the operation of the vehicle" (art, 34-bis, letter b, Vienna Convention) gives the legislators of the individual member States the right to operate a reform intervention which, by introducing into the national systems provisions compliant with the principle expressed in the Convention, will allow level 3 cars—already available in the state of the art—to express their full potential on roads intended for public traffic.[3]

With particular reference to the Italian legal system, the need to amend art. 46 of the Highway Code, the current text of which states that "vehicles mean all machines of any kind circulating on the roads and driven by a human". This profound transformation will bring our system closer to

[1]From this point of view, the classification of the five levels of automation created by the Society of Automotive Engineers International (a standardisation body in the field of the aerospace, automotive and vehicle industry that develops technical standards functional to the manufacture of products and the performance of activities) assumes fundamental importance.. It contemplates a level 0, which designates the traditional car, where driving is entirely devolved to the driver; a level 1, which includes devices (Stability Control; Cruise Control; Lane Correction Technology) now widespread in the vast majority of more recent vehicles; a level 2, which designates vehicles where the automatic devices control both the steering functions and those of acceleration and deceleration in the context of predetermined scenarios, but do not allow the permanent need to maintain constant control by the human driver to be excluded. The subsequent levels of automation, although fully achieved from a technical point of view, concern vehicles that are currently going through - in the overall panorama of a plurality of regulations characterised by differentiated disciplines - the phase of experimentation, both in contexts expressly designed for this purpose and in the conditions of road traffic realities. Thus, level 3 cars have mechanisms capable of controlling every aspect connected to driving in mapped environments, while at the same time requiring the constant presence on board of a driver who carries out monitoring activities and is able to intervene in the event that the scenario prefigured in the mapping changes, necessitating operations that could not be planned on the basis of the information available ex ante. Level 4 of automation includes vehicles in which the presence of a human driver is completely superfluous in all conditions; the presence of tools that allow driving by a person, therefore, is justified only by reason of the eventuality that the car has to be used in environments and conditions in which automated driving is impossible or if the driver decides to dedicate himself personally to driving. The last level (5) includes cars that depart completely from the current concept of vehicle, performing driving functions in a completely autonomous way, excluding any human intervention and which, therefore, are devoid of instruments (steering wheel, pedals) through which this typically occurs.

[2]*Automated Driving Roadmap*, http://www.ertrac.org/uploads/images/ERTRAC_Automated_Driving_2017.pdf.; Ferrari, *Analisi comparata in tema di responsabilità civile legata alla circolazione dei veicoli a guida autonoma*, in Smart Roads *e* driverless cars*: tra diritto, tecnologie, etica pubblica*, edited by Scagliarini, Torino, 2019, p. 99; Tampieri, *L'intelligenza artificiale: una nuova sfida anche per le automobili*, in *Contr. e impr.*, 2020, p. 732; Abraham, Rabin, *Automated Vehicles and Manufacturer Responsibility for Accidents: A New Legal Regime for a New Era* (April 9, 2018), in *Virginia Law Review*, Vol. 105, 2019; Stanford Public Law Working Paper, SSRN: *https://ssrn.com/abstract=3159525*, p. 44; Geistfeld, *A Roadmap for Autonomous Vehicles: State Tort Liability, Automobile Insurance, and Federal Safety Regulation*, in 105 *Calif. L. Rev.* 1611 (2018), p. 1635. Albanese, *La responsabilità civile per i danni da circolazione di veicoli ad elevata automazione*, in *Eur. e dir. priv.*, 2019, p. 995.

[3]Cf. Ilková, Ilka, *Legal aspects of autonomous vehicles - An overview*, in *21st International Conference on Process Control (PC)*, 2017, pp. 428 ss.

those of many US legal systems and more generally to those which, by not adhering to the Vienna Convention, have already been able to admit the generalised circulation of highly automated vehicles for some time.[4]

The opening of the legislator towards forms of circulation in which driving can be largely entrusted to automated systems and requires only an eventual intervention by the driver constitutes a piece of a more articulated design in which another intervention by the legislator of the European Union which—in addition to immediately introducing significant innovations applicable to vehicles equipped with level 2 devices—constitutes an important step towards the diffusion of level 3 cars.

In fact, Reg. (EU) 2019/2144 that entered into force on 6 July 2022—underlining the need to enhance the use of new technologies in order to "further improve the safety performance of vehicles as part of an integrated approach to road safety and in order to better protect vulnerable road users"—has established the obligation to equip newly registered vehicles with a series of functional ADAS devices to increase their active and passive safety (advanced intelligent speed adaptation systems; interface for installing alcohol-type devices; warning of driver inattention and tiredness; advanced driver distraction warning; emergency stop signal; reversing detection) and to monitor the driving phases through the introduction of the so-called Event Data Recorder (art. 6, letter g), Reg. 2019/2144/EU).

The synergistic operation of functional tools to increase vehicle safety and of the tool for recording events (the so-called Event Data Recorder) will make it possible to differentiate newly homologated and registered cars not only in terms of greater safety, but also in terms of the increased ability to reconstruct the dynamics of claims in a timely and potentially unequivocal manner.

The combination of the recent changes to the legislative framework can be observed as the manifestation of a progressive transition from the traditional driving model, entirely entrusted to the action of the driver, to assisted and partially automated driving in which the system governs the driving of the vehicle. The introduction by Reg. (EU) 2019/2144 of new ADAS that assist the driver in driving operations and affect his conduct by inducing him to comply with the rules governing road safety, together with the Event Data Recorder, which records the phases prior to the accidents, enabling accurate reconstruction of their dynamics, make it possible to extend automation level 2 in a general way, and, in other words, to introduce systems capable of limiting human errors and channelling the driver's conduct within compliance with safety standards that traditionally relied on his skills and discipline.

The recent integration of the Vienna Convention through the insertion of the new art. 34-bis, by removing the provision which required the constant involvement of the driver in driving, creates the conditions for the individual systems adhering to the Convention to establish a regulatory framework favourable to the spread of vehicles equipped with level 3 automation devices, i.e. capable of circulating in autonomous mode within mapped environments, requiring driver intervention only in the event that the automated system is unable to operate reliably autonomously.

Ultimately, observing the two provisions as a whole, one can glimpse the transition from a system in which the compliance of driving with the rules governing road safety was entirely entrusted to the conduct of the driver to one in which the discretion and the consequent possibilities of errors in traditional driving are significantly contained by the operation of level 2 ADAS—mandatory installed on newly registered vehicles—and by the gradual progress of vehicles equipped with level 3 automation devices, which will determine a gradual affirmation of automated driving whose safety appears to be guaranteed by the operation of a system programmed to fully comply with road traffic rules. The transition from level 2 to level

[4] JAC WISMANS, SELPI, THYNELL, LINDBERG, *Economics of Road Safety—What does it imply under the 2030 Agenda for Sustainable Development?*, in https://www.uncrd.or.jp/content/documents/4903Final%20background%20paper%20for%20EST%20Plenary%20Session%2010.pdf.

3 can be observed precisely from the perspective of the centrality of the figure of the human driver, who persists in level 2, albeit assisted by the ADAS devices to step down, in level 3, to the role of "supervisor", called upon to intervene in cases where the automated system requires the involvement of the person.[5]

## 2 Containment of Road Accidents and Automation of Driving Between Supranational Principles and National Strategies

The need to guarantee effective protection of the integrity of people against the risks deriving from the circulation of vehicles has always been an objective pursued globally by the United Nations and is included, in the current context, in the Objectives of the UN 2030 Agenda.[6] With this in mind, the Improving Road Safety resolution[7] identifies the objective of limiting the number of deaths and injuries resulting from road traffic by at least 50% by 2030. The aim of limiting the number of accidents and their consequences referring to both car occupants and so-called weak road users, i.e. pedestrians and cyclists, also echoes the programmatic lines that characterise the action of the European Union and that of the national legislator.

In such a scenario, the opportunity to overcome the obstacles posed by the traditional concept of driving and to create the conditions for allowing a rapid, reliable and safe transition towards the diffusion of automated guided vehicles has been highlighted by the European Commission which, within the so-called GEAR 2030, outlined a *Roadmap for Automated Driving,*[8] and by a subsequent report commissioned by the European Parliament in which the opportunity was underlined for a systematic review of many regulations currently in force and the introduction of rules conceived according to the specificities that characterise the new risks arising from the circulation of vehicles, identifying in it the driving force of economic growth[9]; growth which, indeed, will necessarily have to involve a vast network of relationships involving multiple sectors connected to road traffic as summarised in the *Platform for the Deployment of Cooperative Intelligent Transport Systems in the European Union* (C-ITS).[10]

The path undertaken by the European Union, therefore, is aimed at identifying uniform rules capable of promoting the experimentation and subsequent diffusion of highly automated cars in a harmonised legal framework functional to reducing as far as possible the divergences of individual national legislations in all contexts that contribute to regulating the circulation of vehicles, so as to guarantee the reliability of circulation in the multiple situations in which the decision-making traditionally attributable to human beings will be the result of an interaction between human beings and the automated mech-

[5]Ferrari, *Analisi comparata in tema di responsabilità civile legata alla circolazione dei veicoli a guida autonoma*, cit., p. 99; Tampieri, *L'intelligenza artificiale: una nuova sfida anche per le automobili,* cit., p. 732; Di Rosa, *Il* legal framework *internazionale ed europeo*, cit., p. 65.

[6]Jac Wismans, Selpi, Thynell, Lindberg, *Economics of Road Safety—What does it imply under the 2030 Agenda for Sustainable Development?*, cit.

[7]https://documents-dds-ny.un.org/doc/UNDOC/GEN/N20/226/30/PDF/N2022630.pdf?OpenElement.

[8]*Prioritising the Safety Potential of Automated Driving in Europe*, published by the *European Transport Safety Council* (ETSC), in *https://etsc.eu/.*

[9]In the document A common EU approach to liability rules and insurance for connected and autonomous vehicles European Added Value Assessment: Accompanying the European Parliament's legislative own-initiative report (Rapporteur: Mady Delvaux), at https://op.europa.eu/it/publication- detail/-/publication/df658667-20f1-11e8-ac73-01aa75ed71a1, p. 5, it is clarified that the emergence of tools suitable for raising the level of automation of driving will have to lead to a revision of the current regulatory framework and, in particular, of the discipline that governs insurance profiles (Directive 2009/103/EC) and the one concerning the responsibility of the manufacturer (dir. 85/374/CEE).

[10]*Study on the Deployment of C-ITS in Europe: Final Report*, in *https://transport.ec.europa.eu.*

anisms, to which they may sometimes be totally devolved.

Moreover, these considerations fully correspond to the current National Road Safety Plan 2030, which identifies the increase in road safety as one of the pillars of the sustainable mobility system. Precisely in this context, the fundamental importance of ADAS devices is further underlined, which determine a reconciliation of human conduct within the safety standards established by the legislator and monitor the effectiveness of this process through the *Event Data Recorder.*[11] From this point of view, it is necessary to further underline the opportunity for the dissemination of tools capable of increasing the level of compliance of vehicle driving with the rules governing road traffic safety through the containment of that subjectivity factor summed up by the expression "human error". The introduction of a package of ADAS which bring the driving of the vehicle within legal standards significantly limits the possibility of errors; at the same time the possibility of entrusting all driving operations to automated systems subject only to the supervision of the manager contributes to further raising the level of standardisation of vehicle circulation, and consequently compliance with legal disciplines. Nonetheless, it should be emphasised that even in a context intended to contain the incidence of human error, the role of the person who assumes the role of driver will continue to be of fundamental importance, since in the level 2 driving scenario, the action of the driver—although circumscribed within the limitations imposed by the functioning of the ADAS—remains the predominant guiding element; even in the second scenario, i.e. level 3 driving, human conduct retains a fundamental relevance, since the activation of the automated driving mechanisms is in any case left to an operation decided by the driver, which requires the driver to be continuously available and therefore his/her ability to always be ready to intervene appropriately in order to avoid accidents.

[11] *https://www.mit.gov.it/node/15908.*

## 3 The Code Discipline and the Persistent Topicality of Art. 2054, Paragraph 4

The prospect of a constant diffusion of vehicles equipped with devices that will allow access to level 2 and level 3 driving raises delicate questions of interpretation with reference to the persistent relevance of the principles and rules dictated for the governance of the consequences of accidents caused by road traffic.

A first order of issues—which currently arises with reference to level 2 automated vehicles—consists in the need to remodulate the relationship between driver's liability and manufacturer's liability in accidents caused by the vehicle in situations in which the automated devices operate driver assistance (ADAS).[12] The problem of a transition from a driver-focused system to one in which the functions performed by automated mechanisms will assume increasing importance—general and common to all legal systems—arises with specific regard to the national context with a view to verifying the persistent adequacy of the discipline originally conceived by the legislator in 1942 and still in force, which provides that the owner of the vehicle (or, in his place, the usufructuary or the purchaser with retention of title agreement) and the driver are "responsible for damages deriving from construction or lack of maintenance of the vehicle" (art. 2054, paragraph 4, of the Civil Code).[13] To

[12] The problem, analysed in the European context as early as 2005 (VAN WEES, BROOKHUIS, Product liability for ADAS; legal and human factors perspectives, EJTIR, 5, no. 4 (2005), p. 357), has recently shown growing relevance with reference to the US legal system (cf. ABRAHAM, RABIN, Automated Vehicles and Manufacturer Responsibility for Accidents: A New Legal Regime for a New Era, cit., p. 8), which highlights the need to cope with a significant transition period in which the persistent prevalence of traditionally driven cars will lead to maintaining the current rules governing civil liability unchanged; rules which—according to the authors—will have to undergo a profound revision only when the circulation of automated vehicles will reach a measure equal to 25% of the circulating fleet.

[13] ALPA, *Costruzione di autoveicoli, clausole di esonero e responsabilità dell'impresa. Per una diversa lettura dell'articolo 2054, ultimo comma, codice civile*, cit.,

the liability of these subjects, that of the manufacturer must be added. Until the introduction of the discipline on product damage, the manufacturer responded on the basis of the general rule codified in art. 2043 of the Civil Code and that—following the radical change, deriving from the transposition of the dir. 85/374/EEC by d.p.r. no. 224 of 1988—is subject to a liability regime based on the concept of "defective product". In the current regulatory context, therefore, the manufacturer is liable for damage caused to the user on the basis of the rules contained in articles 114 ff. of the consumer code; this form of liability assumes particular importance with regard to damages suffered by the driver of the vehicle, who, until the introduction of the legislation on product damage, could obtain compensation only by resorting to the generic protection offered by art. 2043 of the Civil Code.[14] The growing diffusion of cars equipped with devices that allow assisted driving attributable to level 2 determines a situation in which the product performs driving functions traditionally performed by the driver, thus posing the problem of a transition from a model of responsibility (driver focused) which allocates the compensation cost deriving from the accident to a product-focused one on the latter, which places this cost on the manufacturer of the devices on which the driver may rely. In this perspective, for example, the accident in which a pedestrian is damaged during a parking manoeuvre carried out using advanced systems such as intelligent speed adaptation systems [art. 6, letter. a), EU Reg. 2019/2144] could go beyond the scope of application of art. 2054, paragraph 1, of the Italian Civil Code to fall under the various provisions that base the liability on the product defect by placing the compensation costs that arise from this on the vehicle owner (art. 2054, paragraph 4, of the Civil Code)[15] or of its manufacturer (art. 114 c. cons.). Similar considerations could be repeated with reference to accidents caused during the operation of devices such as the *Automated Lane Keeping System (ALKS)*[16] that conditions the trajectory of the vehicle and stops the driver carrying out a manoeuvre in order to avoid an accident. The analysis of these problems carried out in a detailed study conducted with reference to the German system has highlighted firstly that, if the automated devices presuppose constant interaction with the driver, the latter remains aware of the characteristics of the vehicle, its autonomous driving systems and their limitations. This assumption, in fact, must be considered in order to adopt the most appropriate conduct in terms of monitoring and intervention in case of need. In this new context, therefore, the skills of the driver continue to assume central importance, even if the pace of automation determines their partial modification and a different and greater involvement of the manufacturer. Since the optimal driving of the vehicle will depend on the driver's ability to interact with automated systems, it seems possible to state that the current rules on driver liability for damage caused by traffic will maintain their validity in general terms,[17] but at the same time they must be read in a perspective of ever greater complementarity with respect to the rules governing the liability of the manufacturer. From this point of view, therefore, the information provided by the manufacturer in the instruction booklet and also that communicated through advertising tools are likely to assume increasing importance; in fact, they must not be formulated

1975, I, c. 751; CASTRONOVO, *La nuova responsabilità civile*, 4ª ed., Milano, 2018, p. 758; FRANZONI, *Dei fatti illeciti. Artt. 2043–2059*, in *Comm. cod. civ. Scialoja-Branca-Galgano*, edited by De Nova, Bologna-Roma, 2020, p. 465.

14 AL MUREDEN, *Sicurezza "ragionevole" degli autoveicoli e responsabilità del produttore nell'ordinamento italiano e negli Stati Uniti*, in *Contr. e impr.*, 2012, p. 1505.

15 COLOMBA, Driverless cars *e intelligenza artificiale. Una questione di ordine pubblico: la liceità del brevetto*, in Smart Roads *e* driverless cars*: tra diritto, tecnologie, etica pubblica*, cit., p. 87.

16 *Addendum* 156—UN Regulation No. 157 *Date of entry into force as an annex to the 1958 Agreement: 22 January 2021 Uniform provisions concerning the approval of vehicles with regard to Automated Lane Keeping Systems (ECE/TRANS/WP.29/2020/81).*

17 ABRAHAM, RABIN, *Automated Vehicles and Manufacturer Responsibility for Accidents: A New Legal Regime for a New Era*, cit., p. 5.

in such a way as to induce the driver to overestimate the performance of the automated system and to lower the level of attention required by driving.

These considerations seem likely to retain their validity also with reference to the circulation of cars equipped with level 3 automation mechanisms. In particular, when systems capable of replacing the driver and driving the vehicle autonomously will be operational, it seems likely that the liability for damage caused by the vehicle falls within the provisions of article 2054, paragraph 4, of the Civil Code. However, this does not exclude the persistence of a significant responsibility of the driver who, although not directly involved in driving, must continue to remain alert and ready to intervene as soon as the system requires it. From this point of view, delicate questions arise concerning the functioning of the devices which act as the human machine interface, i.e. that maintain constant communication between the vehicle and the driver in order to promptly and effectively recall the attention of the latter and induce him to take control of the car.

## 4 The Responsibility of the Manufacturer Between Consolidated Paradigms and New Perspectives

The combined implementation of level 2 assisted driving and level 3 partially automated driving should lead, according to the most accredited studies, to a significant reduction in the number of accidents. In this context, however, claims will persist due to vehicle operating defects, imperfections in the road structure or anomalous behaviour by people. However, another portion of accidents could be caused by compliant vehicles, i.e. operating in autonomous mode according to predetermined protocols validated by the legislative standards that define their safety. The first type of accidents, i.e. those caused by defects in the product or infrastructure, as well as abnormal human conduct, could continue to be regulated by resorting to the rules currently governing the liability of the manufacturer, that of the road manager and that of road users .

On the other hand, the second type of claims—i.e. accidents caused by vehicles that comply with legislative standards—poses the problem of allocating costs arising from activities carried out in full compliance with the regulatory standards which define safety levels assessed as reasonable by the legislator.

Thus, the problem arises of creating a regulatory framework capable of providing a sufficient level of certainty and predictability regarding the responsibilities deriving from the marketing and circulation of highly automated vehicles and an efficient allocation of the costs of accidents caused by their circulation. With this in mind, various interpretative solutions have been proposed in the US system in a *de iure condendo* perspective, which is now probably close to materialising.

A vision strongly in favour of achieving a drastic reduction in accidents in the shortest possible time by replacing traditional vehicles with automated ones has identified an interpretative solution based on a coordinated reading of the federal regulations on vehicle safety with the state regulations concerning civil liability. In this perspective, the pre-emption doctrine plays a fundamental role according to which the conformity of a vehicle to the federal standards outlined by the National Highway Traffic Safety Agency (NHTSA) constitutes a limit beyond which a civil liability cannot be configured for the manufacturer who has placed a compliant vehicle on the market.[18] Said principle—formed on the basis

[18]McGarity, *The Preemption War: When Federal Bureaucracies Trump Local Juries*, New Haven London, 2008; Untereiner, *The preemption defense in tort actions: Law Strategy and Practice* Washington D.C., 2008; O'Reilly, *Federal Preemption of State and Local Law*, cit.; Dinh, *Reassessing the law of Preemption*, 88 Geo L.J. 2085 (2000); Goldsmith, *Statutory Foreign Affairs Preemption*, Sup. Ct. Rev. 175 (2000); Sharkey, *Products Liability Preemption: an Institutional Approach*, 76 Geo. Wash. L. Rev., 449 (2008); Davis, *On Restating Products Liability Preemption*, 74 Brooklyn L. Rev. 759 (2009).

of the constant interpretation of the National Traffic and Motor Vehicle Safety Act of 1966 with reference to traditional vehicles—should, according to this vision, also extend to future federal legislation on highly automated vehicles.

Ultimately, the above reading would make it possible to consider regulatory compliance as a complete defence and to promote the rapid diffusion of a technology capable of guaranteeing a higher level of security. In such a perspective, the manufacturer's liability could only arise in the presence of non-compliance with federal standards. In this way, the regulatory-compliant autonomous vehicle should involve a manufacturer's liability limited only to accidents caused by the malfunction of the car and its hardware (strict product liability) or by the absence of adequate warnings provided by the manufacturer in order to allow a reasonably safe use of the vehicle under normal conditions.[19]

In such a design, the extension of the manufacturer's civil liability would not undergo significant changes compared to the current context and the cost of accidents caused by automated vehicles compliant with federal standards would continue to weigh mainly on the user, who would bear the compensation charges deriving from the damages caused to third parties by virtue of his capacity as owner or driver of the vehicle. Also with regard to the so-called one car collision—in which, as a result of a manoeuvre correctly carried out by a compliant automated vehicle, damages were caused to the owner-driver—the pre-emption doctrine would operate as a cause of limitation of the manufacturer's liability and should therefore in principle exclude compensation of the damage suffered by the user of the compliant vehicle.

Such a reading appears in line with the solutions that currently govern liability for damage from compliant products in many product contexts.[20]

This approach could initially appear favourable for manufacturers, as it would limit their liability to only cases of damage caused by defective products, i.e. products that do not comply with legislative standards. Nonetheless, this solution may turn out to be counterproductive as it is not suitable for instilling the feeling of trust in the consumer necessary to overcome the distrust fuelled by the so-called *status quo* bias.[21]

A different reading of the phenomenon of the spread of driverless cars highlights the need to socialise the costs of new accidents caused by the automation of road traffic and to this end highlights the introduction of a system called Market Enterprise Responsibility (MER), which would presuppose the establishment of a fund fed by the manufacturers themselves to create the resources to draw on in order to face compensation for any damage caused. In this way, indemnity coverage would be broadly achieved in favour of the occupants, pedestrians and all types of bystanders, which would allow allocation of all the costs deriving from accidents caused by driverless vehicles. The Market Enterprise Responsibility (MER) system would also make it possible to allocate the costs of accidents caused by automated vehicles according to a method characterised by a high level of economic sustainability by combining with the market share liability mechanism. With this formula, which allows for the distribution of civil liability and the consequent compensation obligations in proportion to the market shares attributable to the individual tortfeasors, it would in fact be possible to place a contribution burden on each manufacturer to finance the fund for compensation of victims of accidents directly proportional to the market shares achieved. The contribution provided by each manufacturer would thus be proportional to the extent of the risk introduced into society quantified by the number of vehicles sold. However, such a crite-

[19] GEISTFELD, *A Roadmap for Autonomous Vehicles: State Tort Liability, Automobile Insurance, and Federal Safety Regulation*, in 105 *Calif. L. Rev.* 1611 (2018), p. 1674.

[20] AL MUREDEN, Product safety *e* product liability *nella prospettiva del danno da prodotto conforme*, in *La responsabilità del produttore*, edited by Alpa, Milano, 2019, p. 489.

[21] CALABRESI & AL MUREDEN, Driverless cars. *Intelligenza artificiale futuro della mobilità*, Bologna, 2021, p. 123.

rion should be supplemented by a further variable that would make it possible to value the choices of companies that invest in improving the safety requirements of their vehicles. In fact, the amount of the contribution due from each manufacturer in order to feed the resources for victim compensation, should be determined by combining the market share parameter with the "safety coefficient" of the vehicles marketed. The latter criterion would constitute an effective incentive to produce and market cars characterised by high capacities to avoid accidents (active safety) and to mitigate their consequences (passive safety).

The prospect of an indemnity-type system in which the resources necessary to ensure prompt and adequate economic compensation to the victims are fed by a contribution from the manufacturers, justified on the basis of market shares achieved and also of the product safety factor, which allows one to understand the fundamental role played by the Event Data Recorder. In fact, the data obtained through the use of this tool, observed as a whole, will allow precise ratings to be made regarding the safety of individual models and, consequently, will create the conditions for being able to reward manufacturers who, by investing in the research and manufacture of vehicles characterised by a high level of safety, may be called upon to contribute to the fund intended for accident victims to an extent inversely proportional to the capacity of their vehicles to avoid accidents and limit their harmful consequences.

Such an approach follows a shared tendency to value decision-making based on the availability and analysis of a significant amount of data (data driven decisions). To this end, therefore, it appears essential to make the relationship of synergy and complementarity between the various disciplines even more intense by ensuring that the devices responsible for data collection are regulated and designed on the basis of increasingly functional data collection to reconstruct every detail of the complex relationships between the driver, assistance devices and driving automation (*Human Machine Interface*).[22]

[22]With reference to the Event Data Recorder (art. 6, letter g), Reg. 2019/2144/EU) there is a reference to reg. UN ECE R 160 which indicates the parameters that must be recorded, together with technical parameters such as recording period, data resolution and accuracy, sampling frequency, and minimum interval. Although this detailed list contains a series of parameters concerning the operation of ABS, airbags, seat belt pretensioners, service brake activation, pitch and roll angles, steering angle, as well as anthropometric data relating to the driver and passengers, limited to body size and weight, no reference is made to any other relevant parameters such as those relating to third party vehicles or infrastructure, rear passengers or pedestrians. Furthermore, data on on-board diagnostics such as those relating to tyre pressure and temperature, condition of the engine and timing components, mechanical deformations, etc. do not appear. To these observations, it must be added that the recording period indicated by the regulatory framework allows the observation of a time window of only 10 s polarised on the collision: the last 5 s before the collision and the first 5 thereafter. In the light of these observations, therefore, further development of the Event Data Recorder system appears desirable, which allows for the observation of a greater quantity of parameters, for recording more complex interactions and, ultimately, for allowing an interaction between technique and legislative discipline based on a relationship of effective synergy.

MIX
Papier aus verantwortungsvollen Quellen
Paper from responsible sources
FSC® C105338

If you have any concerns about our products,
you can contact us on
**ProductSafety@springernature.com**

In case Publisher is established outside the EU,
the EU authorized representative is:
**Springer Nature Customer Service Center GmbH**
**Europaplatz 3, 69115 Heidelberg, Germany**

Printed by Libri Plureos GmbH
in Hamburg, Germany